Anca-Voichita Matioc

Modelling and analysis of nonnecrotic tumors

Anca-Voichita Matioc

Modelling and analysis of nonnecrotic tumors

A Functional Analytic Approach

Südwestdeutscher Verlag für Hochschulschriften

Impressum/Imprint (nur für Deutschland/ only for Germany)
Bibliografische Information der Deutschen Nationalbibliothek: Die Deutsche Nationalbibliothek verzeichnet diese Publikation in der Deutschen Nationalbibliografie; detaillierte bibliografische Daten sind im Internet über http://dnb.d-nb.de abrufbar.

 Alle in diesem Buch genannten Marken und Produktnamen unterliegen warenzeichen-, marken- oder patentrechtlichem Schutz bzw. sind Warenzeichen oder eingetragene Warenzeichen der jeweiligen Inhaber. Die Wiedergabe von Marken, Produktnamen, Gebrauchsnamen, Handelsnamen, Warenbezeichnungen u.s.w. in diesem Werk berechtigt auch ohne besondere Kennzeichnung nicht zu der Annahme, dass solche Namen im Sinne der Warenzeichen- und Markenschutzgesetzgebung als frei zu betrachten wären und daher von jedermann benutzt werden dürften.

Verlag: Südwestdeutscher Verlag für Hochschulschriften Aktiengesellschaft & Co. KG
Dudweiler Landstr. 99, 66123 Saarbrücken, Deutschland
Telefon +49 681 37 20 271-1, Telefax +49 681 37 20 271-0
Email: info@svh-verlag.de
Zugl.: Hanover, Gottfried Wilhelm Leibniz University Hanover, PhD Thesis, 2009

Herstellung in Deutschland:
Schaltungsdienst Lange o.H.G., Berlin
Books on Demand GmbH, Norderstedt
Reha GmbH, Saarbrücken
Amazon Distribution GmbH, Leipzig
ISBN: 978-3-8381-1324-1

Imprint (only for USA, GB)
Bibliographic information published by the Deutsche Nationalbibliothek: The Deutsche Nationalbibliothek lists this publication in the Deutsche Nationalbibliografie; detailed bibliographic data are available in the Internet at http://dnb.d-nb.de.

 Any brand names and product names mentioned in this book are subject to trademark, brand or patent protection and are trademarks or registered trademarks of their respective holders. The use of brand names, product names, common names, trade names, product descriptions etc. even without a particular marking in this works is in no way to be construed to mean that such names may be regarded as unrestricted in respect of trademark and brand protection legislation and could thus be used by anyone.

Publisher: Südwestdeutscher Verlag für Hochschulschriften Aktiengesellschaft & Co. KG
Dudweiler Landstr. 99, 66123 Saarbrücken, Germany
Phone +49 681 37 20 271-1, Fax +49 681 37 20 271-0
Email: info@svh-verlag.de

Printed in the U.S.A.
Printed in the U.K. by (see last page)
ISBN: 978-3-8381-1324-1

Copyright © 2010 by the author and Südwestdeutscher Verlag für Hochschulschriften Aktiengesellschaft & Co. KG and licensors
All rights reserved. Saarbrücken 2010

Prologue

This work is the thesis I wrote during my three years PhD scholarship in the DFG Research Training Group 615 "Interaction of Modeling, Computation Methods and Software Concepts for Scientific-Technological Problems" at the Gottfried Wilhelm Leibniz University of Hanover. For this reason I would like to thank first to all my colleagues from the Leibniz University of Hanover for the nice cooperation and the friendly relationship we had during this years.

Above all, I would like to express my gratitude to my advisor, Prof. Dr. Joachim Escher for offering me the chance to make my PhD study in Hanover. I am grateful for his steady support, encouragement, and for the fruitful discussions and invaluable hints during the last three years.

I am greatly indebted to Prof. Dr. Gerhard Starke for refereeing this thesis and to Prof. Dr. Gheorghe Constantin for guiding my study towards Saarbrücken and Hanover. Especially, I thank my husband Bogdan for his love, encouragement, and for carefully reading the thesis, and my parents for their support and patience. Finally, I am thankful to the Südwestdeutschen Verlag für Hochschulschriften for the nice and efficient collaboration.

Hanover, December 2009   Anca-Voichiţa Matioc

Contents

1 **Introduction**   1

2 **The mathematical model**   13

3 **The radially symmetric case**   19
    3.1 Radially symmetric stationary solutions 21
        3.1.1 The semilinear Dirichlet problem 22
        3.1.2 A representation of ψ_R 29
    3.2 Radially symmetric evolution of tumors 36

4 **The general case**   39
    4.1 Spaces of Hölder continuous functions 39
    4.2 Preparations and the well-posedness result 40
    4.3 The transformed problem 43
    4.4 The nonlinear Cauchy problem 54
    4.5 Besov spaces and Fourier multiplier operators 61
    4.6 Local well-posedness 70

5 **Stability properties**   76
    5.1 Determining the spectrum of the Fréchet derivative of Φ 77
    5.2 Estimates for the symbol of the derivative of Φ 96
    5.3 Instability for $\mathbf{G} > \mathbf{G}_*$ 101
    5.4 Exponential stability for $\mathbf{G} = 0$ 102
    5.5 Exponential convergence of $2\pi/l$–periodic data 109

6 **Local bifurcation analysis**   114

Bibliography   121

Chapter 1

Introduction

Cancer modelling has grown immensely as one of the challenging topics involving applied mathematicians working with researchers active in the biological sciences. The motivation is not only scientific, as in the industrial nations cancer has now moved from seventh to second place in the league table of fatal diseases, being surpassed only by cardiovascular diseases. But what is cancer? Following [32] cancer is characterised by the following two properties:

- reproduces at a faster rate than normal cells,
- invades and continues to proliferate in regions normally occupied by other cells, a process called *metastasis*.

Cancer is classified by the tissue from which they arise and by the type of cells involved. For example, *sarcoma* is a cancer arising in muscles and connective tissue, *leukemia* is a cancer of white blood cells, and *carcinoma* is a cancer originating from epithelial cells, that is, the closely packed cells which align the internal cavities of the body.

Neoplasm, or *tumor*, is a growing mass of abnormal cells. As long as this mass remains clustered together and confined cavity, the tumor is said to be *benign*. If the tumor has emerged out of the cavity, by breaking out through the basal membrane and then proliferating into the extracellular matrix, then the tumor become *malignant*, and we refer to it as *cancer*. When cancer cells invade into the lymphatic vessels or the blood stream they may then be transported into another location and creating there a secondary tumor, the metastasis. The primary tumor (the tumor in the initial location) is traced to a single mutated cell, from which, over a period of time, a colony of cells is formed. A solid tumor may typically be detected only when it reaches a size of 1 cm; by then the tumor contains 10^9 cells, including normal cells. Many gene mutations take place in the human body over one's lifetime. There is evidence that a single abnormal cell, which gives rise to tumor, has risen through a number of genetic mutations. There are two ways by which a gene can become abnormal:

- a stimulating gene becomes hyperactive, or upregulated; such an abnormal gene is called *oncogene*
- an inhibitory gene becomes inactive, or downregulated; it is called a *tumor suppressive* gene (an example being the *p53* gene which controls the initiation of the cell cycle).

The importance of examining the genetic mutations in cancer development is emphasised in [42]. In this paper, the authors identify six critical changes on cell physiology that characterise malignant cancer growth: self-sufficiency in growth signals, insensitivity to anti-growth signal, evading apoptosis, limitless replicative potential, tissue invasion and sustained angiogenesis (the process of formation of new blood vessels, induced by factors secreted by the tumor, and vital for continued tumor growth) and metastasis. All these changes incorporate some aspect of genetic mutation and evolutionary selection leading to malignant progression. The mathematical models have mainly concentrated on only two of these hallmarks: angiogenesis and invasion. Some models have incorporated other hallmarks, without examining the mutations or environments leading to such phenotypes. Although a greater understanding of the biological pathways that lead to these physiological changes will help to create more realistic models, there are still many general questions that can be examined through modelling.

Modelling and simulation of tumor growth is certainly one of the challenging frontiers of applied mathematics which could have a great impact both on the quality of life and the development of the mathematical sciences. Mathematics alone cannot solve the problem of cancer, but applied mathematics may be able to provide a framework in which experimental results can be interpreted, and a quantitative analysis of external actions to control neoplastic growth can be developed. Specifically, models and simulations can reduce the amount of experimentation necessary for drug and therapy development. Moreover, the mathematical theory developed might not only provide a detailed description of the spatiotemporal evolution of the system, but may also help us to understand and manipulate aspects of the process that are difficult to access experimentally.

During the last four decades an increasing number of mathematical models have been proposed to describe the growth of solid tumors (see [10, 21, 22, 35] and the literature therein). There is a three level approach in modelling the complex phenomena influencing and describing the processes inside a tumor: processes on the cellular scale, which are triggered by signals stemming from the sub-cellular level and have an impact on the macroscopic scale, i.e. on the organism, when tumors grow and spread. Very often models combine aspects from this scales.

Models at *sub-cellular level* take into consideration that the evolution of a cell is determined by the genes in its nucleus. Receptors on the cell surface can receive signals which are then transmitted to the cell nucleus, where the genes can be activated or suppressed. In extreme situations, particular signals can induce uncontrolled cell proliferation, or cell death (apoptosis). Unregulated proliferation may activate interactions between tumors cells and host cells, which occur at the cellular level, but are mediated by subcellular processes, such as through signal cascades and receptor expression. These interactions can result in temporary or permanent alterations in gene expression, which in turn can affect a cell's state, such as activation or inactivation of immune cells.

Models at the *cellular level* are proposed to simulate the effects of cell-cell interaction. These interactions are key elements at all stages of tumor formation, whether they are among tumor cells and host cell, or among tumor cells themselves. For example, early in tumor development, if the immune system is active and able to recognise tumor cells, it may be able to develop a destruction mechanism and induce cancer cell death; otherwise, the tumor may evade apoptosis or co-opt the host cell, allowing progressive growth. During invasion and metastasis, alterations in cell-cell adhesion between individual tumor cells are key to driving the process. These cellular interactions are regulated by signals emitted and received by cells through complex transduction processes. Therefore, the connection to the sub-cellular level is evident. On the other hand, the growth of tumor cells will form a mass, so that macroscopic feature become important.

At the *macroscopic level*, the tumor is considered to consist of three zones: an external proliferating zone near high concentration of nutrient, an intermediate layer and an internal zone consisting of necrotic cells only. Prior to vascularisation, these avascular tumors reach an equilibrium size of about 2mm in diameter, where their growth is limited by diffusion of nutrients until the onset of angiogenesis. Although the angiogenic process is often described macroscopically, it occurs through processes at the cellular scale, such as migration, proliferation and cell-cell signaling.

Different mathematical methods correspond to the different scales described above. For instance, models at the cellular scale are generally developed in terms of ordinary differential equations or Boolean networks, while multicellular systems are modeled by nonlinear integro-differential equations similar to those of nonlinear kinetic theory (the Boltzmann equation), or by partial differential equations for systems with internal structures. Macroscopic model lead to systems of nonlinear partial differential equations or discrete modelling approaches.

These different scales are needed to represent the phenomena occurring during tumor growth and all the scales are needed to understand various biological phenomena. Hence, multiscale methods should be developed and the following strategy is one possibility: identify the mathematical structures needed to describe biological phenomena at each scale, then connect the various structures to model the overall system, viewing it as a network of several interconnected subsystems or modules with feed-back down and feed-forward up scales.

A vast literature has been devoted to models which link the cellular scale to the macroscopic tissue scale. In this way, models can address how changes in cell-cell interactions affect the macroscopic properties of the tumor. Once cells have formed a tumor mass, features on the macroscopic level of the tumor environment must be considered. Conversely, the tumor cells themselves can create areas of acidosis, which in turn affect the properties of the cellular and acellular components of the surrounding environment. Nevertheless, there has been much success in using macroscopic models to examine tumor malignancy. The two main types of models used are *continuum*, which examine average behaviour of the densities of populations or components, and *discrete*, which can track the behaviour of individual cells. These types of models employ a wide variety of mathematical methods, as they can describe phenomenological interaction between cell or mechanical interactions based on measuring stresses and strains of the system. All of these methods make some *a priori* assumptions about cell behaviour: they either assume a cell moves through a process like diffusion, or the cellular components act like an elastic fluid.

Continuum models are generally stated in terms of partial differential equations which describe the evolution in time and space of locally averaged quantities related to the behaviour of cell populations. A large variety of continuum models are derived using mass balance equations for the cellular components and reaction-diffusion equations for the chemical or nutrients. The initial system is composed of mass balance equations for the cellular components, extracellular matrix (ECM), and the extracellular fluid (ECF):

$$\rho_j \left[\frac{\partial \phi_j}{\partial t} + \nabla_x \cdot (\phi_j \mathbf{v}_j) \right] = \Gamma_j(\rho, \phi, c), \quad j = 1, ..., L$$

$$\frac{\partial c_i}{\partial t} + \nabla_x \cdot (c_i \mathbf{v}_l) = \nabla_x \cdot (Q_i(\rho, \phi, c) \nabla c_i) + \Lambda_i(\rho, \phi, c), \quad i = 1, ..., M \quad (1.1)$$

where ϕ_j denotes the concentration of each component, e.g. cells, matrix or fluid, c_i denotes the concentration of the chemicals and nutrients, ρ_j are the mass densities of cellular components, $\mathbf{v}_j$ is the mass velocity of the jth population while $\mathbf{v}_l$ is the velocity of the liquid. Moreover, Γ_j is a source term for the particular component which include production and death terms, Q_i is the diffusion coefficient, $\mathbf{v}_i$ is the velocity of the chemical component and Λ_i is the source term for the particular nutrient or chemical. Therefore, the component equations are coupled to the chemical equations via the source terms. For example, blood vessels might be the source of oxygen, which is consumed by the tumor cells and in turn alters the proliferation or death of the tumor cells. It is important to note that the system (1.1) is not closed or self-consistent, and therefore by itself is not a sufficient model. One needs to determine an equation for the velocity $\mathbf{v}_j$ in order to close the system. This leads to the two main classes of macroscopic models, each defined by the choice of movement term. These classes can be defined as *phenomenological* models and *mechanical* models. Phenomenological models make an assumption about movement ignoring mechanical effects, such that cells (or matrix components) do not move, or they move through any combination of diffusion, chemotaxis or haptotaxis. In contrast, mechanical models use force-balance or momentum-balance interaction to determine how the cell, matrix and fluid components move in response to the physical force involved.

Phenomenological models: Most of the tumor growth models are closed by phenomenological assumption, and an explicit equation for the cellular velocity is written. One common example of a phenomenological closure is to assume cells move down a gradient in cell density, which leads to some type of diffusion equation for cell movement. Hence, $\mathbf{v}_j = -D_j \nabla_x \phi_j$, where in most cases D_j is a positive constant, therefore the movement is simply linear diffusion. Recent work has investigated the more general case of $D_j = D_j(\phi, c)$ leading to a nonlinear diffusion term (see [51] and [54]). In [51], Sherratt uses a nonlinear diffusion model for the interaction between tumor cells and ECM. The model was able to suggest a mechanism by which some types of tumors become encapsulated by highly dense ECM. An interesting extension of this work is used in [52], where the authors model movement by contact inhibition in avascular tumor growth. This model also exhibited the traditional avascular tumor structure of a proliferating rim, quiescent band, and necrotic core in terms of continuous cell densities instead of discrete bands of cell types. Alternatively, phenomenological models can specify biased movement such as chemotaxis (see [17]) or haptotaxis (see [7]-[8]). In [48] Sherratt et al. suggested a combination of ECM degradation by proteases and tumor cell haptotaxis as a mechanism for invasion. They performed a traveling wave analysis on a continuum 1-D ODE model for invasive cells, ECM and protease. Perumpanani and Byrne (see [47]) extended the model to use a combination of diffusion and haptotactic movement, and found that ECM heterogeneity affects invasion.

Mechanical models: In contrast with the phenomenological models, mechanical models close the system by specifying cell movement based on physical forces. These models aim to describe how the mechanical properties of the tumor and surrounding tissue influence tumor growth. The momentum balance equations for the constituents are given by

$$\rho\phi_j \left(\frac{\partial \mathbf{v}_j}{\partial t} + \mathbf{v}_j \cdot \nabla_x \mathbf{v}_j \right) = \mathbf{F}_j[\phi, \mathbf{v}], \quad j = 1, ..., L, \quad (1.2)$$

where $\mathbf{F}_j[\phi, \mathbf{v}]$ is describing the forces on the constituent j. We can express $\mathbf{F}_j$ as follows:

$$\mathbf{F}_j = \nabla_x \cdot \mathbf{T}_j + \phi_j \mathbf{f}_j + \mathbf{m}_j,$$

where $\mathbf{m}_j$ is the interaction force with the other constituents, $\mathbf{T}_j$ is the stress tensor and $\mathbf{f}_j$ is the body forces acting on the jth constituent. The equations (1.2) are coupled to the nutrients and chemicals as in the second expression of (1.1), and one use the same mass-balance equation as in the first expression of the above mentioned system. The model (1.2) requires the specification of the constitutive equations relating the forces determining cell motion to the level of stress and compression. For example, as a cell undergoes mitosis and divides into two cells, the daughter cells generate a pressure which displaces the neighbouring cells, thus leading to an increase in tumor size. When using a mechanical continuum description, there are several different classes of mechanical models depending on whether the cells are assumed to behave like a type of fluid or solid medium. If we assume that the cells behave as a fluid, the simplest constitutive equation for the stress comes from assuming the cells act like elastic liquids: $\mathbf{T}_j = -\Sigma_j \mathbf{I}$, with Σ_j the response of the cells to compression. In some special cases, the assumption of the cells moving as an elastic fluid within a rigid ECM can lead to closure by

Darcy's law, where if, for example, $f_j = 0$ and
$$\mathbf{v}_j = -K\nabla_x \Sigma_j,$$
with K the permeability property of the matrix. This constitutive equation has two interpretations: the first is that the system acts as an over-damped force balance, the second is that the fluid, like cells flow through the rigid ECM akin to porous media flow. Alternatively, the cell-matrix milieu can be hypothesised to be like a viscous fluid, where the stress depends on the viscosity as in [12], [15] and [16]. Another class of models views the tumor tissue as a mixture of cells living in a porous medium medium made of ECM and filled with extracellular liquid, see [38]. In conclusion, continuous models are able to exhibit the general behaviour of tumor growth, angiogenesis and invasion. All these continuum approaches model average behaviour at a population level, and fail to examine phenomena that occur at the single cell level.

Discrete models have, in contrast with the continuum models, the ability to track the behaviour of single cells. Due to biotechnological advances, there is an increasing amount of data available on phenomena at a single cell level which merit inclusion in mathematical models. Most discrete models utilise a combination of discrete cell-based models to represent the behaviour of single cells, and continuous equations to model chemical gradients (see [2] and the literature therein). Individual cells are then tracked on a lattice, where the cells of the lattice correspond to biological cells or each cell is made up of several lattice points, as in Potts models (see [55]). Usually, a discrete model is comprised of a regular rectangular lattice for computational simplicity, but alternative geometries can be chosen. Furthermore, the lattice is usually fixed through time, but a free lattice can be constructed to move as a result of cell proliferation. All the discrete models consider the state of each cell or populations of cells to be characterised by the vector variable $\mathbf{w} = \{\mathbf{x}, \mathbf{v}, \mathbf{u}\}$, where $\mathbf{x}$ is position, $\mathbf{v}$ is velocity and $\mathbf{u}$ is a vector detailing the cell's internal state, which might include information such as age, point in cell cycle, or phenotypic characteristics. This informations would in turn affect the probability of a cell moving, proliferating, or changing state. As in continuum models, assumptions regarding cell movement can be implemented, either by using phenomenological models such as diffusion or haptotaxis, or mechanical models. Any mechanical interaction would depend on the cell's position and velocity. An advantage of the discrete method is the ease of embedding subcellular processes within each biological cell. In particular, a discrete approach is useful when modelling angiogenesis, as it allows modelling at the individual cell and vessel level. In this way, mathematical models can examine how endothelial cells link together and form functional blood vessels, and the precise structure of the vascular network (see [8]). Discrete models have also addressed the importance of visco-elastic effects and cell adhesion, with a focus on tumor invasion. Although the basics of tumor growth and movement into the surrounding tissue can be examined through continuum models, advances in imaging now allow us to visualise migration of individual cells and it has been suggested that perhaps single cell behaviour, in contrast to mean cell behaviour, might be driving invasion. Therefore, discrete techniques are better suited to represent these aspects of tumor growth.

There are also other types of models: *biological models,* consisting of coupled ODE systems where the variables correspond to some biological properties of an entire population or *moving boundary models* , when the macroscopic description of biological tissues is obtained from continuum mechanics or microscopic description at cellular level.

Biological models: Biological function is statistically distributed across cells. The biological phenomenon generates the so-called heterogeneity related to progression and to immune activation. Coupled ordinary differential equations can be used to model systems of cell populations, where each variable corresponds to a well-defined biological property characteristic of all cells of the same population. These models are formulated by averaging over the space variable and over the biological function expressed by each population so that the state of the system is simply described by the number density of cells within each population. In [41] the authors initiated a systematic development of population dynamics models focused on cancer. This approach has been developed then by various authors (see [3]). The advantage of the above approach is that models are easily tractable, allowing a relatively rapid identification of the parameters characterising the model by suitable comparisons with experimental data. On the other hand, these simplifications omit potentially important phenomena, such as spatial aspects, and heterogeneity among cells of the same population. Systems of partial differential equations can also be used to model large systems of interacting cells whose microscopic state includes internal variables related to biological functions. The internal variable can be the age of the cell as determined by the cell cycle, which has crucial influence on various biological phenomena such as apoptosis, cell division or mutation. Age-structured models describe the evolution of cellular systems for times of the same order of the cell cycle.

Moving boundary models: The macroscopic description of biological tissues can be obtained either by the classical approach of continuum mechanics or by suitable hydrodynamic limits applied to underlying microscopic description at the cellular level. Different mathematical models, namely elliptic, parabolic, hyperbolic, or simply evolution equations, can be obtained corresponding either to different closures of the conservation equations in the continuum mechanical approach, or to different ratios between the mechanical and biological timescales. One of the most relevant applications of continuum models is the study of moving boundary problems, wherein it is assumed that the growth of solid tumors occurs in an environment, where nutrients nurture their development in the face of chemical factors that inhibit growth, while the immune system an macrophages are also involve in the growth arrest (see [1], [17], [19], [20],[26], [33], [34], [39], and the literature therein). The biological system is schematically represented in the following picture:

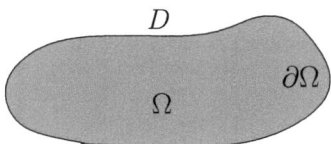

where Ω is the inner zone of the solid tumor, D is the domain of the outer environment and $\partial \Omega$ is the boundary separating Ω from D. To make the model tractable, the geometry is simplified. For example, in the simplest cases Ω is assumed to be spherical, cylindrical ore even one-dimensional.

The next step consists of identifying the macroscopic variables that can possibly describe the state of the overall system viewed as a continuum and then developing a strategy to reduce the complexity related to the large number of components to a limited number of variables. As in [26] the following variables are selected: the density u_1 of living tumor cells, the density u_2 of necrotic tumor cells, the concentration u_3 of the factor that inhibits tumor growth, the concentration u_4 of the factor that activates angiogenesis, the density u_5 of endothelial cells and the concentration u_6 of nutrient. The above variables represent an overall system which is more complex with a greater number of components. The variable u_1 and u_2 are defined in Ω, while u_3, u_4, u_5 and u_6 are defined in the whole domain $\Omega \cup D$.

The mathematical models in the domains Ω and D consist of systems of PDEs derived according to the modelling of the following biological phenomena:

- *Mitosis* occurs only if tumor cells receive a sufficient quantity of nutrient, greater than that needed for survival. On the other hand, division can be inhibited by mitotic inhibitors. When the nutrient levels are too low, cells undergo necrosis.

- *Movement of cells* occurs towards zones of lower cell density, while necrotic tumor cell do not move and naturally disintegrate.

- *Endothelial cells* proliferate with a rate dependent on the chemical factors emitted by tumor cells to activate angiogenesis. Capillary sprouts further facilitate the diffusion of nutrients.

- *Proteins*, named angiostatins, have the ability to reduce the proliferation of endothelial cells and hence reduce angiogenesis.

The system of PDEs is coupled on the free boundary by suitable compatibility conditions, described by ordinary differential equations. In [26], the authors describe in detail how the above biological considerations are converted into a specific mathematical model. In several other papers (see, for example, [19], [23], [33] or [34]) the free and moving boundary problems are studied.

In this thesis we also deal with a moving boundary problem, which is obtained by combining aspects from the cellular and macroscopic scale, and possesses also characteristics of the mechanical model (Darcy's law). Cristini et al. obtained in [19], using algebraic manipulations, a new mathematical formulation of an existing model (see [14, 35, 40]), which describes the evolution of nonnecrotic tumors in different regimes of vascularisation (see the cases $(i) - (iii)$ presented in Chapter 2). The simpler model, as presented in [14, 35, 40], has been studied extensively by different authors, see e.g. [11, 21, 22, 36] and the references therein. In particular, it is shown in these papers that if the parameters belong to an appropriate range, then the mathematical formulation possesses a unique radially symmetric solution. Moreover, the stability properties of this solution under general perturbations, as well as bifurcation phenomena are studied. In contrast, for the more involved model presented in [19] (and which we consider here), not many analytic results are available.

This work is structured as follows: in the second chapter we give a description of the modelling aspects presented in [14, 35, 40]. This model is posed in terms of conservation laws for the nutrient and tumor-cell concentration and contains two parameters $(A, G) \in \mathbb{R}^2$, which have biological meanings. Particularly, varying these constants we find our selfs in different regimes of vascularisation, which correspond to different properties of the tumor dynamics. The mathematical model is a system of PDEs and the variables considered there are the pressure distribution

p inside the tumor, the rate ψ at which nutrient is added to the tumor domain and the boundary of the tumor, which induces a third, more important, variable.

In the third chapter we analyse the radially symmetric case, which was treated in [27]. The analysis done in here will also serve as ancillary tool in the forthcoming chapters. The case $G = 0$ corresponds to a simpler model describing also the motion of a fluid blob in a Hele-Shaw cell (see [29]). This is the case, for example, when the mitosis rate is zero, cf. (2.14). In this situation any circle describes a stationary tumor, and we show in Chapter 5 that the circles near the unit circle $\mathbb{S}$ build a three dimensional manifold which attracts, at an exponential rate, solutions of the system which are initially nearby.

This is no longer the case when $G \neq 0$. Our analysis shows that there exists a (unique) radially symmetric steady state of the tumor provided $A \in (0, f(1))$, where f is the nutrient consumption rate. The radius of this stationary tumor depends only on the parameter A. The behaviour of a tumor being initially a circle centred in 0 is described in Theorem 3.0.3. Particularly, Theorem 3.0.3 shows that if $G > 0$ and $A \in (0, f(1))$, a tumor, which is close to the unique stationary tumor and which is also radially symmetric, exists in the large and converges exponentially towards the stationary state.

This fact may suggest that the unique equilibrium, which exists provided $A \in (0, f(1))$, is stable if $G > 0$. We show that this impression is false by considering in the following chapters arbitrary initial tumor domains. We prove in Theorem 4.2.1 that the problem is locally well-posed in time in the small Hölder spaces $h^{m+\beta}(\mathbb{S})$, $m \in \mathbb{N}$ and $\beta \in (0, 1)$, context. To prove this theorem we transform the system on a fixed domain by using the so-called Hanzawa diffeomorphism. Though, this transformation has the disadvantage of adding nonlinearities, it allows us to transform the problem into an abstract evolution equation on $\mathbb{S}$

$$\partial_t \rho = \Phi(\rho), \qquad \rho(0) = \rho_0,$$

where ρ_0 is the function defining the contour of the tumor at time $t = 0$ and Φ is a nonlocal and nonlinear operator depending smoothly on ρ. We prove that the Fréchet derivative $\partial \Phi(0)$, considered as an unbounded operator in $h^{1+\alpha}(\mathbb{S})$, with dense definition domain $h^{4+\alpha}(\mathbb{S})$, generates a strongly continuous semigroup in $\mathcal{L}(h^{1+\alpha}(\mathbb{S}))$ for all $\alpha \in (0, 1)$. By using the properties of the continuous interpolation functor defined by Grisvard and Da Prato in [24], Theorem 4.2.1 follows from general results for abstract evolution equations as presented in [45].

In Chapter 5 we show that the radially symmetric equilibrium found for $A \in (0, f(1))$ and $G > 0$ is unstable provided G is large enough, cf. Theorem 5.3.1.

That the equilibrium is unstable for $G < 0$ was already establish in Theorem 3.0.3. The proof of Theorem 5.3.1 is obtain after determining the spectrum of the Fréchet derivative $\partial\Phi(0)$. The derivative $\partial\Phi(0)$ is a Fourier multiplier operator and its spectrum consists only of eigenvalues which approach $-\infty$. However, appropriate estimates show that some of this eigenvalues are positive if G is large, which yields the result stated in Theorem 5.3.1. The exponential stability result established in Theorem 3.0.3 (a) is generalised in Theorem 5.5.1. We show that solutions corresponding to $2\pi/l$−periodic initial data, where the positive integer l is related to G, converge exponentially to the radially symmetric equilibrium, provided $G > 0$.

Finally, in Chapter 6, we prove that there exist also other stationary states when the tumor is no longer radially symmetric. This stationary states are obtained as bifurcation branches from the trivial, radially symmetric, solution. The constant G is the bifurcation parameter, and, considering appropriate subspaces of the small Hölder spaces, we can apply the classical result on bifurcations from simple eigenvalues due to Crandall and Rabinowitz to obtain, in Theorem 6.0.6, these special stationary solutions.

Chapter 2

The mathematical model

We introduce first the system we are interested in. As widely used in the modelling, the tumor is treated as an incompressible fluid and tissue elasticity is neglected. Cell-to-cell adhesive forces are modeled by surface tension at the tumor-tissue interface. The growth of the tumor is governed by a balance between cell-mitosis and apoptosis (programed cell-death). The rate of mitosis (λ_M) depends on the concentration of nutrient and no inhibitor chemical species are present.

The two-dimensional system describing the evolution of the tumor is a fully nonlinear system consisting of two decoupled Dirichlet problems, one for the rate ψ at which nutrient is added to the tumor domain $\Omega(t)$ and one for the pressure p inside the tumor. These two variables are coupled by a relation which describes the dynamic of the tumor boundary. The full system reads as follows

$$\begin{cases} \Delta \psi = f(\psi) & \text{in } \Omega(t), \quad t \geq 0, \\ \Delta p = 0 & \text{in } \Omega(t), \quad t \geq 0, \\ \psi = 1 & \text{on } \partial\Omega(t), \quad t \geq 0, \\ p = \kappa_{\partial\Omega(t)} - AG\frac{|x|^2}{4} & \text{on } \partial\Omega(t), \quad t \geq 0, \\ G\frac{\partial \psi}{\partial n} - \frac{\partial p}{\partial n} - AG\frac{n \cdot x}{2} = V(t) & \text{on } \partial\Omega(t), \quad t > 0, \\ \Omega(0) = \Omega_0. \end{cases} \quad (2.1)$$

Here Ω_0 is the initial state of the tumor, V is the normal velocity of the tumor boundary, $\kappa_{\partial\Omega(t)}$ the curvature of $\partial\Omega(t)$, and the constants A and G have biological meaning (see below the explicit formulae). The function $f \in C^\infty([0, \infty))$ has the following properties

$$f(0) = 0 \quad \text{and} \quad f'(\psi) > 0 \quad \text{for} \quad \psi \geq 0. \tag{2.2}$$

In [19] the special case $f(r) = r$ is considered. In this situation, the first equation of the system is linear, and if the tumor domain is a sphere or an infinite cylinder, then the solution is known through an explicit formula. The general case is considerably more involved, the equation being highly nonlinear.

This system was obtained by Cristini and his co-workers, making use of algebraic manipulations, from an existing model (see [14, 35, 40]). In the following we give a description of the old model. As before, let $\Omega(t)$ denote the tumor domain at time $t \geq 0$. The quasi-steady diffusion equation for the nutrient concentration σ is

$$0 = D \cdot \Delta \sigma + \psi, \tag{2.3}$$

where D is the diffusion constant and ψ the rate of change of the nutrient. This rate ψ incorporates all sources (in our situation only blood provides the tumor with nutrient) and sinks of nutrient in the tumor volume. Nutrient is supplied by the vasculature at a rate $\psi_B(\sigma, \sigma_B)$, where σ_B is the nutrient concentration in blood. The rate of consumption of nutrient by the tumor cells is $\lambda \sigma$, where λ is a constant. The blood-tissue transfer is assumed to be linear

$$\psi_B = -\lambda_B (\sigma - \sigma_B), \tag{2.4}$$

where λ_B is a positive constant. Thus, the rate ψ is given by

$$\psi = -\lambda_B (\sigma - \sigma_B) - \lambda \sigma. \tag{2.5}$$

Because the tumor is treated as an incompressible fluid, the velocity field $\mathbf{v}$ in $\Omega(t)$ satisfies the continuity equation

$$\nabla \mathbf{v} = \lambda_P, \tag{2.6}$$

with λ_P the cell-proliferation rate, given by the expression

$$\lambda_P = b\sigma - \lambda_A, \tag{2.7}$$

where b and λ_A are again positive constants. The constant λ_A plays an important role in our analysis because it describes the rate of apoptosis. The velocity is assumed to obey Darcy's law

$$\mathbf{v} = -\mu \nabla P, \tag{2.8}$$

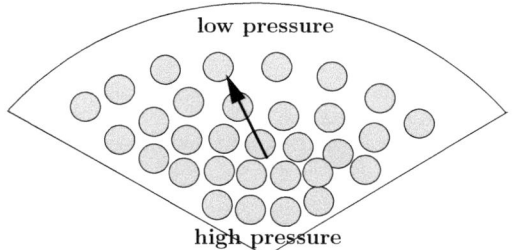

Figure 2.1: Darcy's law

where μ is the cell mobility and P is the pressure inside $\Omega(t)$. This means that the cells move from regions with high pressure to regions where the pressure is low. This principle is illustrated in Figure 2.1. The condition for the concentration at the boundary $\partial\Omega(t)$ is given by

$$\sigma = \sigma^{\infty}, \qquad (2.9)$$

with σ^{∞} the nutrient concentration outside the tumor volume, assumed to be constant. The characteristic mitosis rate is set to be $\lambda_M = b\sigma^{\infty}$. Pressure is assumed to satisfy the Laplace-Young boundary condition

$$P = \gamma \kappa_{\partial\Omega(t)}, \qquad (2.10)$$

where γ is the surface tension related to cell-to-cell adhesive forces. The normal velocity $V = n \cdot \mathbf{v}$ at the tumor boundary $\partial\Omega(t)$ is

$$V = -\mu\, n \cdot \nabla P, \qquad (2.11)$$

with n the outward unit normal at $\partial\Omega(t)$.

Nondimensionalisation: The relations (2.3) and (2.5) reveal that there is an intrinsic length scale

$$L_D = D^{\frac{1}{2}} \cdot (\lambda_B + \lambda)^{-\frac{1}{2}},$$

which, for $\lambda_B = 0$, estimates the stable size of an avascular tumor when diffusion of nutrient and consumption balance. By nondimiensionalising lengths with L_D

we obtain, from (2.8) and (2.10), an intrinsic relaxation time scale λ_R^{-1}, corresponding to the rate
$$\lambda_R = \mu \gamma L_D^{-3}, \tag{2.12}$$
associated to the relaxation mechanism, cell mobility and surface tension. This rate is used to nondimensionalise time.

We introduce now the parameters A and G by setting:
$$A = \frac{\frac{\lambda_A}{\lambda_M} - B}{1 - B}, \tag{2.13}$$

$$G = \frac{\lambda_M}{\lambda_R} \cdot (1 - B). \tag{2.14}$$

The parameter G describes the relative rate of mitosis to the relaxation mechanism (cell mobility and cell-to-cell adhesion). The parameter A describes the balance between apoptosis and mitosis. Both parameters also include the effect of vascularisation B, which is given by
$$B = \frac{\sigma_B}{\sigma^\infty} \frac{\lambda_B}{\lambda + \lambda_B}. \tag{2.15}$$

The parameters A and G allow us to subdivide the growth of the tumor into three regimes associated with increasing degrees of vascularisation:

(i) *low vascularisation* (diffusion dominated, e.g. *in vitro* cell cultures) :

$G \geq 0$ and $A > 0$ (*i.e.* $B < \lambda_A/\lambda_M$),

(ii) *moderate vascularisation* : $G \geq 0$ and $A \leq 0$ (*i.e.* $1 > B \geq \lambda_A/\lambda_M$),

(iii) *high vascularisation* : $G < 0$ (*i.e.* $B > 1$).

The moderate and high vascularisation cases correspond to the regimes observed in *in vivo* experiments.

Further on, we introduce the modified concentration $\overline{\psi}$ and the modified pressure $\overline{p}$ so that:
$$\sigma = \sigma^\infty(1 - (1 - B)(1 - \overline{\psi})), \tag{2.16}$$

$$P = \frac{\gamma}{L_D}\left(\overline{p} + G(1 - \overline{\psi}) + AG\frac{|x|^2}{4}\right). \tag{2.17}$$

The following lemma shows that the problem just stated above can be reformulated in terms of the system (2.1). Indeed, dropping down the bars in (2.18)-(2.20) yields the two-dimensional version of (2.1).

Lemma 2.0.1 *The dimensional problem given by the relations (2.3)–(2.11) is equivalent to the following two non-dimensional decoupled problems:*

$$\begin{cases} \Delta\bar{\psi} = \bar{\psi} & \text{in } \Omega(t), \\ \bar{\psi} = 1 & \text{on } \partial\Omega(t), \end{cases} \tag{2.18}$$

and

$$\begin{cases} \Delta\bar{p} = 0 & \text{in } \Omega(t), \\ \bar{p} = \kappa_{\partial\Omega(t)} - AG\dfrac{|x|^2}{4} & \text{on } \partial\Omega(t). \end{cases} \tag{2.19}$$

The tumor surface evolves with normal velocity

$$V = -n \cdot \nabla\bar{p} + Gn \cdot \nabla\bar{\psi} - AG\dfrac{n \cdot x}{2} \quad \text{on } \partial\Omega(t). \tag{2.20}$$

Proof From the relations (2.3), (2.5), (2.15) and (2.16) we find that

$$\Delta\bar{\psi} = \dfrac{\lambda_B + \lambda}{D}\bar{\psi} \quad \text{in } \Omega(t). \tag{2.21}$$

The balance between diffusion and nutrient consumption ($L_D = 1$) yields the first equation of (2.18). Equation (2.16) further implies

$$\bar{\psi} = \dfrac{\sigma - \sigma^\infty B}{\sigma^\infty(1 - B)},$$

and, in view of (2.9), we obtain the following boundary condition on $\partial\Omega(t)$

$$\bar{\psi} = \dfrac{\sigma^\infty - \sigma^\infty B}{\sigma^\infty - \sigma^\infty B} = 1.$$

From the relations (2.6)-(2.8) we obtain that

$$\Delta P = \dfrac{1}{\mu}(\lambda_A - b\sigma).$$

Recalling $\bar{\psi} = (\sigma - \sigma^\infty B)/(\sigma^\infty(1 - B))$, the relations (2.13), (2.14), (2.16), and (2.17) yield

$$\Delta\bar{p} = \dfrac{1}{\mu\gamma}(\lambda_A - b\sigma) + \dfrac{\sigma - \sigma^\infty B}{\sigma^\infty(1 - B)}\dfrac{\lambda_M}{\lambda_R}(1 - B) - \dfrac{\lambda_M}{\lambda_R}(1 - B)\dfrac{\frac{\lambda_A}{\lambda_M} - B}{1 - B}.$$

Furthermore, (2.12) implies
$$\Delta \bar{p} = \frac{\lambda_A}{\lambda_R} - \frac{\lambda_A}{\lambda_R} - \sigma \left(\frac{b}{\lambda_R} - \frac{\lambda_M}{\lambda_R \sigma^\infty} \right).$$

Since $b = \lambda_M/\sigma^\infty$, it follows that
$$\Delta \bar{p} = 0 \quad \text{in} \quad \Omega(t).$$

The second equation of (2.19) follows now from (2.10) and (2.17).

Equation (2.20) follows straightforwardly from previous relations. This completes the proof. □

Chapter 3

The radially symmetric case

This chapter is dedicated to the radially symmetric case. In Section 3.1 we transform the system describing the radially symmetric solutions of (2.1) into an equation for the radius R. This equation will be solved using a nice representation of the function ψ (see formula (3.31)). The main results of this chapter, Theorem 3.0.2 and Theorem 3.0.3, will be proved in Section 3.2 by making use of ODE–techniques.

Our analysis shows that, under the assumptions (2.2), problem (2.1) possesses a unique radially symmetric solution iff $(A, G) \in (0, f(1)) \times (\mathbb{R} \setminus \{0\})$. The radius of this tumor, which we denote by R_A, depends only on the constant A. This matches also the previous results obtained for the simpler model (see [21, 35]). Furthermore, given $(A, G) \in \mathbb{R}^2$, we show that the tumor exists in the large, provided it is radially symmetric initially. The asymptotic behaviour of these tumors for $t \to \infty$ is described in Theorem 3.0.3.

If $G = 0$ then any disc is a stationary solution. Therefore, we consider only the case $G \neq 0$ here. We give now the main results of this chapter:

Theorem 3.0.2 (Steady states) *Given $(A, G) \in (0, f(1)) \times (\mathbb{R} \setminus \{0\})$, there exists a unique radially symmetric, stationary solution to problem (2.1). The radius R_A of the stationary tumor depends only on the parameter A, and decreases with respect to this variable.*

Given $R > 0$, we denote by $D(0, R)$ the disc in $\mathbb{R}^2$ with centre 0 and radius R.

Theorem 3.0.3 (Asymptotic behaviour) *If the tumor is initially radially symmetric, i.e. $\Omega_0 = D(0, R_0)$ for some $R_0 > 0$, then it remains symmetric for* all times and the following assertions hold:

(a) If $(A, G) \in (0, f(1)) \times (0, \infty)$, there exist positive constants $\omega, C > 0$ and r such that if $|R_0 - R_A| \leq r$, then

$$|R(t) - R_A| \leq Ce^{-\omega t} \quad \text{for} \quad t \geq 0. \tag{3.1}$$

(b) If $(A, G) \in [f(1), \infty) \times (0, \infty)$, we have

$$R(t) \searrow_{t \to \infty} 0. \tag{3.2}$$

(c) If $(A, G) \in [f(1), \infty) \times (-\infty, 0)$, we have that

$$R(t) \nearrow_{t \to \infty} \infty. \tag{3.3}$$

(d) If $(A, G) \in (0, f(1)) \times (-\infty, 0)$, then the stationary solution is unstable, since

$$R(t) \xrightarrow[t \to \infty]{} \begin{cases} 0, & \text{if } R_0 < R_A, \\ \infty, & \text{if } R_0 > R_A. \end{cases} \tag{3.4}$$

(e) If $(A, G) \in (-\infty, 0] \times (-\infty, 0) \cup (0, \infty)$, then

$$R(t) \xrightarrow[t \to \infty]{} \begin{cases} 0, & \text{if } G < 0, \\ \infty, & \text{if } G > 0. \end{cases} \tag{3.5}$$

In all the above situations (a)–(e) the convergence is monotone. Moreover, $R(t)$ converges exponentially fast, except the case (b), when $f(1) = A$ and in the case (e), when $A = 0$ and $G > 0$.

Compared with [35], where a vascularised tumor has been studied, our model here includes also the case of low and moderate vascularised regimes. In the low vascularised regime it is possible that the tumor disappears if the parameter A is large enough, meaning that sufficiently many cells are dying. Furthermore, also in the high vascularised regime it is possible that the tumor vanishes, but in this case the parameter A must be very small. Notice that, if initially tumor growth occurs, then the radius of the tumor is always increasing in time. Also, if the radius decreases at $t = 0$, then the tumor shrinks as time evolves.

3.1 Radially symmetric stationary solutions

Let $(A, G) \in \mathbb{R}^2$ be given. In the radially symmetric case, i.e. for $\Omega(t) = D(0, R(t))$, system (2.1) reads as follows

$$\begin{cases} \Delta \psi = f(\psi) & \text{in } D(0, R(t)), \quad t \geq 0, \\ \Delta p = 0 & \text{in } D(0, R(t)), \quad t \geq 0, \\ \psi = 1 & \text{on } \partial D(0, R(t)), \quad t \geq 0, \\ p = \dfrac{1}{R(t)} - AG\dfrac{R^2(t)}{4} & \text{on } \partial D(0, R(t)), \quad t \geq 0, \\ G\dfrac{\partial \psi}{\partial n} - \dfrac{\partial p}{\partial n} - AG\dfrac{R(t)}{2} = R'(t) & \text{on } \partial D(0, R(t)), \quad t > 0, \\ R(0) = R_0, \end{cases} \quad (3.6)$$

where $n = x/|x|$ and $R_0 > 0$. Notice that if $G = 0$, then the function ψ is eliminated from the kinematic boundary condition and, therefore, any disc is a stationary solution of (3.6) (see the discussion at the beginning of Section 5.4). In the remaining of this section we consider therefore just the case $G \neq 0$.

We stress that the boundary conditions in (3.6) depend only on the radius of the tumor. This motivates us to look for solutions $(R, \psi, p) \in (0, \infty) \times C^2([0, R]) \times C^2([0, R])$, solving the coupled problem

$$\begin{cases} \psi'' + \dfrac{1}{r}\psi' = f(\psi) & 0 < r < R, \\ p'' + \dfrac{1}{r}p' = 0 & 0 < r < R, \\ \psi(R) = 1 & \\ p(R) = \dfrac{1}{R} - AG\dfrac{R^2}{4} & \\ G\psi'(R) - p'(R) - \dfrac{AGR}{2} = 0. & \end{cases} \quad (3.7)$$

The system (3.7) has been obtained from (3.6), by representing the Laplacian Δ in polar coordinates. Problems (3.6) and (3.7) are equivalent in the following sens: if (R, ψ, p) is a solution of (3.7) then, letting $\Omega_0 = D(0, R)$, the triple $(\Omega_0, \psi(|x|), p(|x|))$ is a radially symmetric stationary solution of (3.6). Conversely, it is also immediate that any stationary solution of (3.6) determines a unique solution of (3.7).

3.1.1 The semilinear Dirichlet problem

We prove first that, given $R > 0$, the semilinear Dirichlet problem, consisting of the first two equations of (3.6), possesses a unique solution. This is equivalent to showing that the system

$$\begin{cases} \psi'' + \dfrac{1}{r}\psi' = f(\psi), & 0 < r < R, \\ \psi(R) = 1, \end{cases} \quad (3.8)$$

possesses a unique solution $\psi \in C^2([0, R])$. To this scope we shall use the so called *shooting method*. Suppose we have found a solution to (3.8). Multiplying with r the differential equation of (3.8) we get that $(r\psi'(r))' = rf(\psi(r))$ for $r \in [0, R]$. Integrating, we obtain

$$\psi'(r) = \dfrac{1}{r}\int_0^r sf(\psi(s))\,ds. \quad (3.9)$$

Thus, we have $\psi'(0) = 0$. Therefore, given $c \in (0, 1)$, consider the initial value problem

$$\begin{cases} \psi'' + \dfrac{1}{r}\psi' = f(\psi), & 0 < r, \\ \psi(0) = c, \\ \psi'(0) = 0. \end{cases} \quad (3.10)$$

Theorem 3.1.1 *Given $c \in (0, 1)$, there exists a unique maximal solution $\psi_c \in C^\infty([0, R_c^*))$ of (3.10) and*

$$\lim_{r \nearrow R_c^*} \psi_c(r) = \infty. \quad (3.11)$$

Proof We prove first short time existence and uniqueness of the solution to (3.10). From (3.9) we obtain that any solution to (3.10) satisfies

$$\psi(r) = c + \int_0^r \dfrac{1}{s}\int_0^s tf(\psi(t))\,dt\,ds, \quad r \geq 0. \quad (3.12)$$

Let $b > 0$, $M := \max_{[0, c+b]}(f + f')$ and $h := \min\{\sqrt{b/M}, \sqrt{1/2M}\}$. The set

$$B := \{u : [0, h] \to \mathbb{R}_+ \mid u \text{ is continuos}, \|u - c\|_\infty \leq b\}$$

is a closed set in the Banach space $(C([0, h], \mathbb{R}), \|\cdot\|_\infty)$, thus is a complete metric space. The mapping $T : B \to B$, defined by

$$Tu(r) := c + \int_0^r \frac{1}{s} \int_0^s t f(u(t))\, dt\, ds \quad \text{for} \quad r \geq 0 \text{ and } u \in B,$$

is $1/2$-contraction. Indeed, for $u \in B$ we have

$$|Tu(r) - c| = \int_0^r \frac{1}{s} \int_0^s t f(u(t))\, dt\, ds \leq \int_0^r \int_0^s f(u(t))\, dt\, ds \leq Mh^2 \leq b$$

for all $r \in [0, h]$, thus $T(B) \subset B$. Moreover, for $u, v \in B$, we have

$$|Tu(r) - Tv(r)| \leq \int_0^r \frac{1}{s} \int_0^s t |f(u(t)) - f(v(t))|\, dt\, ds \leq Mh^2 \|u - v\|_\infty$$

for all $r \in [0, h]$, hence $\|Tu - Tv\|_\infty \leq 1/2 \|u - v\|_\infty$. The Banach fixed point theorem implies that T has a unique fixed point ψ_c in B, thus (3.10) has a unique solution in the set $C^2([0, h])$.

Let R_c^* denote the maximal existence time for this solution. It follows easily that ψ_c is the unique solution to (3.10) satisfying $\psi_c(0) = c$. Using an induction argument, we get that $\psi_c \in C^\infty([0, R_c^*))$.

We are left to prove relation (3.11). We first observe that ψ_c and ψ_c' are both strictly increasing on $[0, R_c^*)$. Indeed, given $0 < t < R_c^*$, we see that ψ_c' is a solution of the problem

$$\begin{cases} u'' + \dfrac{1}{r} u' - \left(\dfrac{1}{r^2} + f'(\psi_c) \right) u = 0, & 0 < r < t, \\ u(0) = 0. \end{cases} \quad (3.13)$$

Because ψ_c' is positive and continuous on $[0, t]$, it must attain its maximum in t. Let h be the constant found in the first part of the proof. The pair (ψ_c, ψ_c') is the unique solution of the problem

$$\begin{cases} (u, v)' = g(r, (u, v)), & \dfrac{h}{2} < r < R_c^*, \\ (u, v)\left(\dfrac{h}{2}\right) = \left(\psi_c\left(\dfrac{h}{2}\right), \psi_c'\left(\dfrac{h}{2}\right) \right), \end{cases} \quad (3.14)$$

with $g : (0, \infty) \times \mathbb{R}^2 \to \mathbb{R}^2$ given by $g(r, (u, v)) = (v, -v/r + f(u))$. We conclude that $R_c^* = \infty$ or $|(\psi_c, \psi_c')| \to \infty$ as $r \to R_c^*$.

Suppose now that $R_c^* = \infty$. Since $\psi_c \geq c$, we obtain for $r \geq 0$ that

$$\psi_c'(r) = \frac{1}{r}\int_0^r sf(\psi_c(s))\,ds \geq \frac{1}{r}\int_0^r sf(c)\,ds \geq \frac{r^2}{2r}f(c) \geq \frac{r}{2}f(c).$$

A further integration shows that $\lim_{r \to R_c^*} \psi_c(r) = \infty$. In the other case we have either $\psi_c \to \infty$ or $\psi_c' \to \infty$, and the conclusion holds.

$\square$

Given $c \in (0,1)$, let $\psi_c \in C^\infty([0, R_c^*))$ be the solution of problem (3.10). There exists a unique $R_c \in (0, R_c^*)$ such that

$$\psi_c(R_c) = 1.$$

Further on we study the mapping $[c \mapsto R_c]$.

Lemma 3.1.2 *Let $\mathcal{R} : (0,1) \to \mathbb{R}_+$ be given by $\mathcal{R}(c) = R_c$. We observe that for $c_1 < c_2$ we have $\psi_{c_1} < \psi_{c_2}$, $R_{c_1}^* \geq R_{c_2}^*$, and $R_{c_1} > R_{c_2}$, i.e. $\mathcal{R}$ is a strictly decreasing function.*

As the next results show, the function $\mathcal{R}$ is continuous and bijective.

Lemma 3.1.3 *Given $c_0 \in (0,1)$ and $R < R_{c_0}^*$, we have*

$$\lim_{c \nearrow c_0} \max_{[0,R]} |\psi_c - \psi_{c_0}| = 0. \tag{3.15}$$

Proof Since $c < c_0$ we have $R_c^* \geq R_{c_0}^*$, $\psi_c \in C^\infty([0, R_{c_0}^*))$, and $\psi_c < \psi_{c_0}$. Let $R < R_{c_0}^*$ be given, and set $M := \max_{[0,\psi_{c_0}(R)]} f'$. Given $r \in [0, R]$, we get

$$|\psi_{c_0}(r) - \psi_c(r)| \leq |c_0 - c| + \int_0^r \frac{1}{t}\int_0^t s|f(\psi_{c_0}(s)) - f(\psi_c(s))|\,ds\,dt$$

$$\leq |c_0 - c| + \int_0^r \int_0^t M|\psi_{c_0}(s) - \psi_c(s)|\,ds\,dt$$

$$\leq |c_0 - c| + MR\int_0^r |\psi_{c_0}(s) - \psi_c(s)|\,ds.$$

Gronwall's lemma implies

$$|\psi_{c_0}(r) - \psi_c(r)| \leq (c_0 - c)e^{MR^2},$$

an the proof is completed.

$\square$

As a direct consequence of the previous lemma we obtain:

Corollary 3.1.4 *The function $\mathcal{R}$ is continuous from the left.*

Proof Let $c_0 \in (0,1)$ be given, and presuppose that $\lim_{c \nearrow c_0} \mathcal{R}(c) > \mathcal{R}(c_0)$. We find then a constant $b \in (\mathcal{R}(c_0), R^*_{c_0})$ such that $\mathcal{R}(c_0) < b < \mathcal{R}(c)$ for all $c \in (0, c_0)$. Lemma 2.2 yields
$$1 \geq \lim_{c \nearrow c_0} \psi_c(b) = \psi_{c_0}(b) > 1,$$
and we are done.

$\square$

Proposition 3.1.5 *The function $\mathcal{R}$ is continuous.*

Proof We are left to prove that $\mathcal{R}$ is continuous from the right. We assume by contradiction that $\lim_{c \searrow c_0} \mathcal{R}(c) < \mathcal{R}(c_0)$. Given $c \in (0,1)$, let $\widetilde{R}_c \in (0, \infty)$ be defined as the point with $\psi_c(\widetilde{R}_c) = 2$. The mapping $[c \mapsto \widetilde{R}_c]$ is also strictly decreasing. From (3.12) we obtain
$$\psi_c(\widetilde{R}_c) - \psi_c(R_c) = \int_{R_c}^{\widetilde{R}_c} \frac{1}{r} \int_0^r t f(\psi_c(t)) \, dt \, dr,$$
hence
$$\begin{aligned} 1 &\leq (\widetilde{R}_c - R_c) \max_{[0, \widetilde{R}_c]} \frac{1}{r} \int_0^r t f(\psi_c(t)) \, dt \\ &\leq (\widetilde{R}_c - R_c) f(2) \widetilde{R}_{c_0} \end{aligned}$$
for all $c > c_0$. Thus, we get
$$\lim_{c \searrow c_0} \widetilde{R}_c \geq \frac{1}{f(2) \widetilde{R}_{c_0}} + \lim_{c \searrow c_0} R_c.$$

Choose $a > 0$, such that

$$\min\left\{\frac{1}{f(2)\widetilde{R}_{c_0}} + \lim_{c\searrow c_0} R_c, \mathcal{R}(c_0)\right\} > a > \lim_{c\searrow c_0} R_c.$$

Since $\lim_{c\searrow c_0} \widetilde{R}_c > a$ there exists $c_1 > c_0$ with $\widetilde{R}_c > a$ if $c \in [c_0, c_1]$. Consequently, $[0, a] \subset [0, \widetilde{R}_c] \subset [0, R_c^*]$ for all $c_0 \leq c \leq c_1$. Letting

$$M := \max_{[0, \psi_{c_1}(a)]} f'$$

we repeat the arguments presented in the proof of Lemma 3.1.3 to obtain

$$|\psi_c(r) - \psi_{c_0}(r)| \leq M(c - c_0)e^{Ma^2}$$

for all $c \in [c_0, c_1]$ and $r \in [0, a]$. It follows that $\psi_c(a) - \psi_{c_0}(a) \to_{c\searrow c_0} 0$, and so

$$1 > \psi_{c_0}(a) = \lim_{c\searrow c_0} \psi_c(a) \geq 1.$$

This is a contradiction, thus our assumption was false, and the proof is complete. $\square$

Proposition 3.1.6 *The mapping* $\mathcal{R} : (0, 1) \to (0, \infty)$ *is bijective. More precisely, we have*

$$\lim_{c\nearrow 1} \mathcal{R}(c) = 0 \quad \text{and} \quad \lim_{c\searrow 0} \mathcal{R}(c) = \infty. \tag{3.16}$$

Proof Let $c \in (0, 1)$ be given. From (3.9) we obtain that $\psi_c'(r) \geq f(1/2)\, r/2$ for all $0 \leq r < R_c^*$, provided $c > 1/2$. Integrating, we get

$$1 = \psi_c(R_c) \geq c + \frac{R_c^2}{4} f\left(\frac{1}{2}\right),$$

and letting $c \nearrow 1$, we obtain the first relation in (3.16).

We presuppose now that $\lim_{c\searrow 0} R_c =: T < \infty$. Following the lines in the proof of Proposition 3.1.5 we have

$$\psi_c(\widetilde{R}_c) - \psi_c(R_c) \leq (\widetilde{R}_c - R_c)\widetilde{R}_c f(2). \tag{3.17}$$

Integrating the relation $(r\psi_c'(r))' = rf(\psi_c(r))$, we obtain

$$\psi_c'(r) = \frac{R_c \psi_c'(R_c)}{r} + \frac{1}{r}\int_{R_c}^r sf(\psi_c(s))\,ds$$

for all $R_c \leq r < R_c^*$. Thus

$$\begin{aligned}
1 = \psi_c(\widetilde{R}_c) - \psi_c(R_c) &= R_c \psi_c'(R_c)\int_{R_c}^{\widetilde{R}_c}\frac{1}{r}\,dr + \int_{R_c}^{\widetilde{R}_c}\frac{1}{r}\int_{R_c}^r sf(\psi_c(s))\,ds\,dr \\
&\geq f(1)(\widetilde{R}_c - R_c)^2/4
\end{aligned}$$

for all $c \in (0,1)$. Consequently,

$$\widetilde{R}_c \leq \frac{2}{\sqrt{f(1)}} + R_c \leq \frac{2}{\sqrt{f(1)}} + T. \tag{3.18}$$

The relations (3.17) and (3.18) enforce

$$\widetilde{R}_c \geq R_c + \frac{1}{f(2)\left(\frac{2}{\sqrt{f(1)}} + T\right)}.$$

Since $R_c \nearrow T$ for $c \searrow 0$, there exists $c_0 \in (0,1)$ with $\widetilde{R}_c > T$ for all $c \in (0, c_0]$, and so $[0,T] \subset [0, R_c^*)$ for $c \in (0, c_0]$. Further on, we have $\psi_c(r) \leq \psi_{c_0}(r) \leq \psi_{c_0}(T)$ for all $r \in [0,T]$ and $c \in (0, c_0]$. With $M := \max_{[0,\psi_{c_0}(T)]} f'$ we obtain

$$\psi_c'(r) \leq \int_0^r (f(\psi_c(\tau)) - f(0))\,d\tau \leq M\int_0^r \psi_c(\tau)\,d\tau$$

for all $r \in [0,T]$ and $c \in (0, c_0]$. Integrating and using Fubini's theorem we get

$$\psi_c(r) \leq c + MT\int_0^r \psi_c(s)\,ds$$

for $r \in [0,T]$ and $c \leq c_0$. Gronwall's lemma implies

$$\|\psi_c\|_{\infty,[0,T]} \leq ce^{MT^2}, \;\forall c \in (0, c_0]. \tag{3.19}$$

Letting $c \searrow 0$ in relation (3.19), we conclude $\|\psi_c\|_{\infty,[0,T]} \to 0$, in contradiction with $R_c < T$ and $\psi_c(R_c) = 1$. Thus, our assumption was false, and this completes the proof. □

With these preparations we come to the following result:

Theorem 3.1.7 *Given $R > 0$, there exists a unique solution $\psi_R \in C^\infty([0, R^*))$, $R < R^*$, to problem (3.8).*

Proof Let $R > 0$ be given. From Proposition 3.1.5 and Proposition 3.1.6 we conclude that, there exists a unique $c_0 \in (0, 1)$ such that $R_{c_0} = R$. Thus, $\psi_R := \psi_{c_0}$ is a solution of (3.8).

We are left to prove that this solution is unique. Let $\psi_R(x) = \psi_R(|x|)$, $x \in \overline{D}(0, 1)$. The function ψ_R is then solution of the problem

$$\begin{cases} \Delta \psi = f(\psi) & \text{in } D(0, R), \\ \psi = 1 & \text{on } \partial D(0, R). \end{cases} \quad (3.20)$$

Let φ be another solution of (3.20). Then $u = \psi_R - \varphi \in C^2(\overline{D}(0, R))$ must solve the Dirichlet problem

$$\begin{cases} \Delta u = gu & \text{in } D(0, R), \\ u = 0 & \text{on } \partial D(0, R), \end{cases}$$

with $g(x) = \int_0^1 f'(t\psi(x) + (1-t)\varphi(x))\, dx$ for all $x \in D(0, R)$. From (2.2) we have $g > 0$ which, together with the weak maximum principle, implies $u = 0$, and we are done. □

When considering the problem for p the situation is simpler, and we obtain that p is constant depending on A, G, and R, as the next lemma shows:

Lemma 3.1.8 *The problem*

$$\begin{cases} p'' + \dfrac{1}{r}p' = 0, & 0 < r < R, \\ p(R) = \dfrac{1}{R} - AG\dfrac{R^2}{4}, \end{cases} \quad (3.21)$$

possesses a unique solution $p_R \in C^\infty([0, \infty))$, given by

$$p_R(r) = \frac{1}{R} - AG\frac{R^2}{4}, \quad (3.22)$$

for all $r \in [0, \infty)$.

Summarising, we have reduced the existence problem for (3.7) to an equation on the real line. More precisely, from Theorem 3.1.7 and Lemma 3.1.8 we get that problem (3.7) has solutions iff the equation

$$A = \frac{2\psi_R'(R)}{R} \tag{3.23}$$

is solvable. However, it is difficult to treat (3.23) directly. We shall first derive a suitable representation of ψ_R.

3.1.2 A representation of ψ_R

In the following we consider the parameter-depending problem

$$\begin{cases} \dfrac{\partial^2 U}{\partial r^2}(r,\lambda) + \dfrac{1}{r}\dfrac{\partial U}{\partial r}(r,\lambda) &= \lambda f(U(r,\lambda)), \quad 0 \le r \le 1, \\ \dfrac{\partial U}{\partial r}(0,\lambda) &= 0, \\ U(1,\lambda) &= 1, \end{cases} \tag{3.24}$$

with λ in $[0,\infty)$. As in the previous subsection we can show that for each $\lambda \in [0,\infty)$ there exists a unique, non-decreasing solution $U(\cdot,\lambda) \in C^\infty([0,1])$ of (3.24).

We prove now that the Fréchet derivative of U is uniformly Lipschitz- continuous on compact subsets of $[0,1] \times [0,\infty)$, and that for fixed $r \in [0,1]$, the partial function $U(r,\,\cdot\,)$ is non-increasing on $[0,\infty)$.

Lemma 3.1.9 *Given $r \in [0,1]$, the mapping $[0,\infty) \ni \lambda \mapsto U(r,\lambda)$ is decreasing.*

Proof If $\lambda = 0$, then $U(r,0) = 1$ for all $r \in [0,1]$. Let $0 < \lambda_1 < \lambda_2$, $U_1 := U(\cdot,\lambda_1)$, $U_2 := U(\cdot,\lambda_2)$, and suppose that $c_1 = U_1(0) < U_2(0) = c_2$, with $c_1, c_2 \in (0,1)$. Then, since $U_1(1) = U_2(1) = 1$, we find $\bar{r}$ with $U_1(\bar{r}) = U_2(\bar{r})$, and $U_1 < U_2$ on $[0,\bar{r}]$. For $t \in [0,\bar{r}]$ we have

$$U_2'(t) - U_1'(t) = \frac{\lambda_2}{t}\int_0^t s f(U_2(s))\,ds - \frac{\lambda_1}{t}\int_0^t s f(U_1(s))\,ds$$

$$\ge \frac{\lambda_2}{t}\int_0^t s\left(f(U_2(s)) - f(U_1(s))\right)ds \ge 0.$$

Integrating, we obtain $U_2(t) - U_1(t) \geq c_2 - c_1 > 0$ on $[0, \bar{r}]$, and that is a contradiction, so $c_2 \leq c_1$.

Suppose now that $c_2 = c_1$. Then we obtain

$$\begin{aligned} U_2'' - U_1'' + \frac{1}{r}(U_2' - U_1') &= \lambda_2 f(U_2) - \lambda_1 f(U_1) \\ &= \lambda_2(f(U_2) - f(U_1)) + (\lambda_2 - \lambda_1)f(U_1), \end{aligned}$$

and using the mean value theorem, we get for $t \in [0, 1]$ that

$$f(U_2(t)) - f(U_1(t)) = \int_0^1 f'(U_1(t) + s(U_2(t) - U_1(t))) \cdot (U_2(t) - U_1(t))\, ds.$$

Letting $W = U_2 - U_1$ we have

$$\begin{cases} W'' + \frac{1}{t}W' - gW > 0, & \text{in } (0,1), \\ W(0) = W(1) = 0, \end{cases} \quad (3.25)$$

with a smooth, positive function g. Therefore, $W < 0$ in $(0, 1)$. Further on we have

$$\begin{aligned} \lim_{t \to 0}(U_2''(t) - U_1''(t)) &= \lim_{t \to 0}\left[-\frac{\lambda_2}{t^2}\int_0^t sf(U_2(s))\, ds \right. \\ &\quad \left. + \frac{\lambda_1}{t^2}\int_0^t sf(U_1(s))\, ds + \lambda_2(f(U_2)) - \lambda_1(f(U_1)) \right] \\ &= \frac{\lambda_2 - \lambda_1}{2}f(c_2) > 0. \end{aligned}$$

Hence, we find $\delta \in (0, 1)$ such that

$$U_2(t) - U_1(t) \geq \int_0^t \frac{\lambda_2 - \lambda_1}{4}f(c_2)s\, ds > 0, \ \forall t \in [0, \delta],$$

in contradiction with $U_2 < U_1$ on $(0, 1)$. Thus, $c_2 < c_1$ for $\lambda_1 < \lambda_2$. The weak maximum principle implies $U_2 < U_1$ in $[0, 1)$. $\square$

For our further analysis we also need the following theorem:

Theorem 3.1.10 *Let X, Y, Z be metric spaces. Then $f : X \times Y \to Z$ is uniformly Lipschitz-continuous iff f is uniformly Lipschitz-continuous with respect to both variables.*

Proof See the proof of [5, Lemma 8.1]. □

Lemma 3.1.11 *U is uniformly Lipschitz-continuous on $[0, 1] \times [0, N]$ for all $N \in (0, \infty)$.*

Proof Let $N > 0$ be fixed. For $(x, \lambda) \in [0, 1] \times [0, \infty)$ we obtain from (3.24) that

$$U(x, \lambda) = 1 - \int_0^1 \frac{1}{r} \int_0^r s\lambda f(U(s, \lambda))\, ds\, dr + \int_0^x \frac{1}{r} \int_0^r s\lambda f(U(s, \lambda))\, ds\, dt. \tag{3.26}$$

Relation (3.26) implies that

$$[0, 1] \ni r \mapsto U(r, \lambda)$$

is uniformly Lipschitz-continuous with respect to $\lambda \in [0, N]$. For $0 \leq \lambda < \mu$ we have $U(\cdot, \lambda) \geq U(\cdot, \mu)$. It follows

$$0 < U(0, \lambda) - U(0, \mu) \leq \int_0^1 \frac{1}{r} \int_0^t s(\mu - \lambda) f(1)\, ds\, dr \leq f(1)(\mu - \lambda).$$

Thus,

$$|U(0, \lambda) - U(0, \mu)| \leq f(1)|\mu - \lambda|. \tag{3.27}$$

From (3.26), for $\lambda, \mu \geq 0$ and letting $M = \max_{[0,1]} f' > 0$, we obtain, from (3.27) and Fubini's theorem, that

$$|U(x, \lambda) - U(x, \mu)| \leq 2f(1)|\mu - \lambda| + NM \int_0^x |U(s, \lambda) - U(s, \mu)|\, ds.$$

The conclusion follows using Gronwall's lemma and Theorem 3.1.10. □

From Lemma 3.1.11 we obtain the following result:

Lemma 3.1.12 *The partial derivative of U with respect to r is uniformly Lipschitz-continuous on $[0, 1] \times [0, N]$ for any fixed $N \in (0, \infty)$.*

Proof Let $N \in (0, \infty)$ be fixed and $V(r, \lambda) := \partial U/\partial r(r, \lambda)$. We have that

$$\frac{V(r,\lambda)}{r} = \frac{1}{r^2}\int_0^r s\lambda f(U(s,\lambda))\,ds \leq \frac{\lambda}{2}f(1)$$

for all $r \in [0, 1]$, and together with (3.24) we get

$$\left|\frac{\partial V}{\partial r}(r,\lambda)\right| \leq \frac{3}{2}\lambda f(1) \leq \frac{3}{2}Nf(1) := L, \; \forall (r,\lambda) \in [0,1] \times [0,N].$$

Consequently, $|V(r, \lambda) - V(s, \lambda)| \leq L|r - s|$. Thus, $[0, 1] \ni r \mapsto V(r, \lambda)$ is uniformly Lipschitz-continuous with respect to λ.

For $r \in (0, 1]$ and $\lambda, \mu \geq 0$ we have

$$|V(r,\lambda) - V(r,\mu)| \leq (f(1) + NML_N)|\lambda - \mu|,$$

where L_N is the Lipschitz constant for U on $[0, 1] \times [0, N]$. Using the continuity of V with respect to r, and applying once again Theorem 3.1.10 we obtain the desired result. □

Let us now study the differentiability of U with respect to the variable λ. Given $\lambda \geq 0$, we consider the following Dirichlet problem

$$\begin{cases} \Delta v(x,\lambda) - \lambda f'(U(x,\lambda))v(x,\lambda) = f(U(x,\lambda)) & \text{in } D(0,1), \\ v(x,\lambda) = 0 & \text{on } \partial D(0,1), \end{cases} \quad (3.28)$$

where $U(x, \lambda) := U(|x|, \lambda)$ is the solution of the Dirichlet problem

$$\begin{cases} \Delta U(x,\lambda) = \lambda f(U(x,\lambda)) & \text{in } D(0,1), \\ U(x,\lambda) = 1 & \text{on } \partial D(0,1). \end{cases} \quad (3.29)$$

Problem (3.28) has a unique solution $v \in C^\infty(\overline{D}(0, 1))$ and, using the estimate (3.12) in [37], we find a positive constant c, depending only on f, such that

$$\|v(\cdot, \lambda)\|_{\infty, \overline{D}(0,1)} \leq c\|f(U(\cdot, \lambda))\|_{\infty, \overline{D}(0,1)} \leq cf(1), \; \forall \lambda \geq 0. \quad (3.30)$$

Let now $N > 0$ and $\lambda \in [0, N]$ be fixed. Let further $\mu \in [0, 1]$, if $\lambda = 0$, respectively $\mu \in [-\lambda/2, 1/2]$, if $\lambda \in (0, N]$. We set

$$\omega(x) := U(x, \lambda + \mu) - U(x, \lambda) - \mu v(x, \lambda),$$

for all $|x| \leq 1$. Then $\omega \in C^\infty(\overline{D}(0,1))$ is a solution of the Dirichlet problem

$$\begin{cases} \Delta\,\omega(x) - \lambda f'(U(x,\lambda))\omega(x) = h(x) & \text{in } D(0,1), \\ \omega(x) = 0 & \text{on } \partial D(0,1), \end{cases}$$

where $h \in C^\infty(\overline{D}(0,1))$ satisfies $|h(x)| \leq C|\mu|^2$ for all x in $\overline{D}(0,1)$ and μ in the appropriate set, with a constant C independent of μ. Applying again estimate (3.12) in [37] we obtain

$$\|\omega\|_{\infty, \overline{D}(0,1)} \leq c|\mu|^2.$$

Particularly, U is differentiable on $D(0,1) \times [0,\infty)$ with respect to the variable λ and $\partial U/\partial \lambda = v$.

Lemma 3.1.13 *Given $N \in (0,\infty)$, the mapping $[0, N] \ni \lambda \mapsto v(x,\lambda)$ is uniformly Lipschitz-continuous with respect to the variable $x \in D(0,1)$.*

Proof Let $\lambda, \mu \in [0, N]$ be given and set $w(x) = v(x, \lambda) - v(x, \mu)$ for x in $\overline{D}(0,1)$. Since w is a solution of the Dirichlet problem

$$\begin{cases} \Delta w - \lambda f'(U(\cdot, \lambda))w = f(U(\cdot, \lambda)) - f(U(\cdot, \mu)) + v(\cdot, \mu)[\lambda f'(U(\cdot, \lambda)) \\ \qquad\qquad\qquad\qquad\qquad -\mu f'(U(\cdot, \mu))] & \text{in } D(0,1), \\ w = 0 & \text{on } \partial D(0,1), \end{cases}$$

we get that $\|w\|_{\infty, \overline{D}(0,1)} \leq c\,|\mu - \lambda|$, with a constant c depending only on f and N, cf. (3.30). This completes the proof. $\square$

Given $t \in [0,1]$, $x \in \overline{D}(0,1)$ with $t = |x|$ and $\lambda \in [0, \infty)$, we have

$$\lim_{\mu \to 0} \frac{U(t, \lambda+\mu) - U(t, \lambda)}{\mu} = \lim_{\mu \to 0} \frac{U(|x|, \lambda+\mu) - U(|x|, \lambda)}{\mu}$$

$$= \lim_{\mu \to 0} \frac{U(x, \lambda+\mu) - U(x, \lambda)}{\mu} = v(x, \lambda),$$

thus v is radially symmetric. Lemma 3.1.13 implies that U is differentiable on $[0,1] \times [0,\infty)$ with respect to λ and that $\partial U/\partial \lambda = v$. Of course we consider the restriction of v to the unit interval $[0,1]$ here.

We still have to show that v is uniformly Lipschitz-continuous on $[0,1] \times [0,N]$, $N > 0$, with respect to r. Therefore, it is enough to prove that $\partial v/\partial r$ is uniformly bounded on $[0,1] \times [0,N]$. Indeed, from (3.28), in view of the radial symmetry, v is the solution to the problem

$$\begin{cases} \dfrac{\partial^2 v}{\partial r^2}(r,\lambda) + \dfrac{1}{r}\dfrac{\partial v}{\partial r}(x,\lambda) - \lambda f'(U(r,\lambda))v(r,\lambda) = f(U(r,\lambda)), & 0 < r < 1 \\ v(r,\lambda) = 0, & r = 1. \end{cases}$$

We then obtain

$$\frac{\partial v}{\partial r}(r,\lambda) = \frac{1}{r}\int_0^r s[\lambda f'(U(s,\lambda))v(s,\lambda) + f(U(s,\lambda))]\,ds,$$

and the uniform boundedness of $\partial v/\partial r$ on $[0,1] \times [0,N]$ follows in virtue of (3.30).

Combining the results obtained in Theorem 3.1.10 and Lemmas 3.1.11-3.1.13 yields:

Theorem 3.1.14 *The mapping U is continuously differentiable on $[0,1] \times [0,\infty)$, with uniformly Lipschitz-continuous derivatives on compact subsets of $[0,1] \times [0,\infty)$.*

The function U will play an important role in our analysis. First it will help us to determine the solution of (3.8), as the next theorem shows:

Theorem 3.1.15 *Given $R > 0$ we denote by ψ_R the solution to (3.8). We have*

$$\psi_R(r) = U\left(\frac{r}{R}, R^2\right), \quad 0 \leq r \leq R. \tag{3.31}$$

Proof The proof follows by direct computations.

$\square$

With all these preparation we are now able to give the proof of the first main result of this chapter, the existence and uniqueness of radially symmetric, stationary solution to problem (2.1).

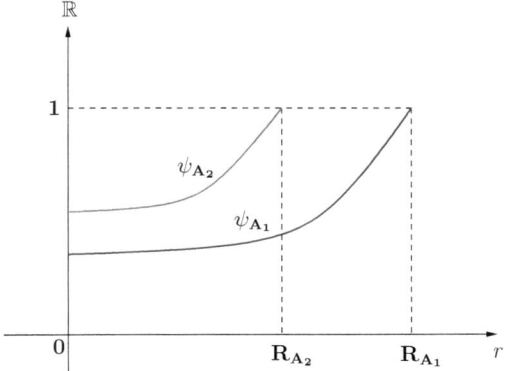

Figure 3.1: The radius R_A and the function ψ_A for $A_1 < A_2$.

Proof *(Proof of Theorem 3.0.2)* From Theorem 3.1.7, Theorem 3.1.15, Lemma 3.1.8, and relation (3.23) we conclude that (3.7) has a solution iff there exists a positive R solving the equation

$$A = \frac{2}{R^2} \frac{\partial U}{\partial r}(1, R^2) = 2 \int_0^1 r f(U(r, R^2)) \, dr. \tag{3.32}$$

We define the function

$$F : [0, \infty) \to \mathbb{R}, \qquad F(R) = 2 \int_0^1 r f(U(r, R^2)) \, dr. \tag{3.33}$$

From Lemma 3.1.9 and Theorem 3.1.14 we get that F is continuously differentiable and strictly decreasing on $[0, \infty)$, with $F(0) = f(1)$.

Moreover, we state that $\lim_{R \to \infty} F(R) = 0$. If we had that $U(\,\cdot\,, \lambda) \to_{\lambda \to \infty} 0$ a.e. on $[0, 1]$, then this would follow by Lebesgue's dominated convergence theorem, since U is bounded by 1. Let us thus presuppose that this is not the case, i.e. there exists $r_0 \in (0, 1)$ such that

$$U(r_0, \lambda) \geq a, \quad \forall \lambda \geq 0,$$

with a positive constant a. It follows then

$$1 \geq U(1,\lambda) - U(r_0,\lambda) \geq \lambda \int_{r_0}^1 \frac{1}{r} \int_{r_0}^r sf(U(s,\lambda))\,ds\,dr$$

$$\geq \lambda \int_{r_0}^1 \frac{r_0}{r} \int_{r_0}^r f(a)\,ds\,dr \geq \lambda r_0 f(a) \frac{(1-r_0)^2}{2} \xrightarrow[\lambda \to \infty]{} \infty.$$

This is a contradiction, thus $U(\,\cdot\,,\lambda) \to_{\lambda \to \infty} 0$ a.e. on $[0,1]$.

Summarising, $F([0,\infty)) = (0, f(1)]$, hence (3.32) has a positive solution iff $A \in (0, f(1))$. Given $A \in (0, f(1))$, let R_A denote the unique solution to (3.32). Lemma 3.1.9 implies then that $(0, f(1)) \ni A \mapsto R_A$ is strictly decreasing, see Figure 3.1. This completes the proof. □

3.2 Radially symmetric evolution of tumors

In this section we aim to prove that problem (3.6) possesses for any initial data $R_0 > 0$ a unique, globally defined solution. A triple $\big(R(\cdot), \psi_{R(\cdot)}, p_{R(\cdot)}\big)$ is called solution to (3.6) if

$$R \in C^1([0,\infty), (0,\infty)),$$
$$(\psi_{R(t)}, p_{R(t)}) \in C^2([0, R(t)]), \quad \text{for} \quad t \geq 0$$

and if $\big(R(\cdot), \psi_{R(\cdot)}, p_{R(\cdot)}\big)$ satisfies the equations in (3.6) pointwise.

From Theorem 3.1.7 and Lemma 3.1.8 we deduce that $\big(R(\cdot), \psi_{R(\cdot)}, p_{R(\cdot)}\big)$ is a solution of the evolution problem (3.6) iff the function R solves the following initial value problem

$$\begin{cases} R' = h(R), & 0 < t < T, \\ R(0) = R_0, \end{cases} \tag{3.34}$$

where $R_0 > 0$ is the radius describing the initial state of the tumor and

$$h(R) := GR \left(\int_0^1 rf(U(r, R^2))\,dr - \frac{A}{2} \right) = \frac{GR}{2}(F(R) - A), \quad R \geq 0,$$

with F defined in (3.33).

The asymptotic behaviour of a tumor, which is initially radially symmetric was described in Theorem 3.0.3. In the following we give the proof of this theorem.

Proof *(Proof of Theorem 3.0.3)* Let $(A, G) \in \mathbb{R}^2$ be fixed. The function h is continuously differentiable, and since $U \leq 1$, has linear growth. For $R_0 > 0$ fixed, we define R to be the maximal defined solution of (3.34). Let T denote the maximal existence time of this solution. It follows that the solution R is bounded on bounded intervals (see [5, Theorem 7.8]). If $G = 0$, then $R = R_0$ on $[0, \infty)$. We consider now the case $G \neq 0$. We split our argumentation as follows:

(a) Let $(A, G) \in (0, f(1)) \times (0, \infty)$. In this case 0 and R_A are the only stationary solutions of (3.34). Since $h > 0$ on $(0, R_A)$ and $h < 0$ on (R_A, ∞) it must hold that $T = \infty$ and $R(t) \to R_A$ as $t \to \infty$. Moreover,

$$h'(R_A) = \frac{GR_A}{2} F'(R_A) = 2GR_A^2 \int_0^1 r f(U(r, R_A^2)) \frac{\partial U}{\partial \lambda}(r, R_A^2) \, dr := -\omega < 0.$$

Hence R_A is exponentially stable.

(b) Let $(A, G) \in [f(1), \infty) \times (0, \infty)$. In this case $R = 0$ is the only zero of h and $h < 0$ on $(0, \infty)$. It follows then $T = \infty$ and $\lim_{t \to \infty} R(t) = 0$. Moreover,

$$\frac{R'}{R} < \frac{G}{2}(f(1) - A) < 0,$$

which implies that $R(t)$ converges exponentially for $A > f(1)$.

(c) Let $(A, G) \in [f(1), \infty) \times (-\infty, 0)$. Since $G > 0$ we have in this situation that $h > 0$ on $(0, \infty)$. It follows that $R(t) \to \infty$ for $t \to T$. Because R must be bounded on bounded intervals we conclude that $T = \infty$. That the convergence is at an exponential rate follows from the following inequality

$$\frac{R'}{R} \geq \frac{G}{2}(F(R_0) - A).$$

(d) Let $(A, G) \in (0, f(1)) \times (-\infty, 0)$. As in (a), 0 and R_A are the only two zeros of h, but now $h < 0$ on $(0, R_A)$ and $h > 0$ on (R_A, ∞). If $R_0 < R_A$, then, as in (b), R decreases in infinite time to 0. In this case the exponential convergence is implied by

$$\frac{G}{2}(F(R) - A) < \frac{G}{2}(F(R_0) - A) < 0.$$

If $R_0 > R_A$ then we have the situation from (c), i.e. $\lim_{t \to \infty} R(t) = \infty$. Since

$$\frac{G}{2}(F(R) - A) > \frac{G}{2}(F(R_0) - A) > 0,$$

the $R(t)$ converges exponentially fast.

(e) In the situation considered here, h and G have the same sign. The radius $R(t)$ increases to infinity if $G > 0$ and shrinks to zero for negative G. Moreover, we have that

$$\frac{R'}{R} \geq -\frac{AG}{2}, \quad \text{if } G > 0,$$

$$\frac{R'}{R} \leq \frac{G}{2}(F(R_0) - A), \quad \text{if } G < 0,$$

on $[0, \infty)$. The desired result follows in view of the above relations. $\square$

Chapter 4

The general case

We leave now the radially symmetric case and focus on the general, i.e. non-symmetric situation, when the tumor domain is a perturbation of a circle $R \cdot \mathbb{S}$, with $R > 0$. Here, $\mathbb{S}$ stands for the unit circle $\mathbb{S} = \{x \in \mathbb{R}^2 : |x| = 1\}$. We first introduce the Banach spaces we shall frequently use in this thesis.

4.1 Spaces of Hölder continuous functions

The functions defined on $\mathbb{S}$ can be identified with $2\pi-$periodic functions on $\mathbb{R}$. Given $m \in \mathbb{N}$, the Banach spaces $C^m(\mathbb{S})$ consist of $2\pi-$periodic and real-valued functions on $\mathbb{R}$, which are $m-$times continuously differentiable. We write

$$\|f\|_{C^m(\mathbb{S})} := \sum_{l=0}^{m} \max_{x \in \mathbb{S}} |f^{(l)}|(x)$$

for the Banach norm on $C^m(\mathbb{S})$. Let now $\beta \in (0, 1)$. The *Hölder space* $C^{m+\beta}(\mathbb{S})$ is the subspace of $C^m(\mathbb{S})$ consisting only of functions with the property that the Hölder seminorm

$$[f]_{m,\beta} := \sup_{\substack{t,s \in \mathbb{R} \\ t \neq s}} \frac{|f^{(m)}(t) - f^{(m)}(s)|}{|t-s|^\beta}$$

is finite. Endowed with the norm

$$\|f\|_{C^{m+\beta}(\mathbb{S})} := \|f\|_{C^m(\mathbb{S})} + [f]_{m,\beta} \quad \text{for} \quad f \in C^{m+\beta}(\mathbb{S}),$$

$(C^{m+\beta}(\mathbb{S}), \|\cdot\|_{C^{m+\beta}(\mathbb{S})})$ is a Banach space.

The Hölder spaces are compactly embedded, $C^s(\mathbb{S}) \hookrightarrow C^r(\mathbb{S})$ for $r < s$. However, the embedding is not dense if $r \notin \mathbb{N}$. If we have to deal with semigroups in the Hölder spaces context it is natural to introduce the so-called *small Hölder spaces* $h^{m+\beta}(\mathbb{S})$, $m \in \mathbb{N}$ and $\beta \in (0,1)$, which are defined as the closure of the smooth functions $C^\infty(\mathbb{S})$ in the $C^{m+\beta}(\mathbb{S})$−norm. Since $C^\infty(\mathbb{S}) \subset h^{m+\beta}(\mathbb{S})$ for all $m \in \mathbb{N}$ and $\beta \in (0,1)$, the embedding

$$h^{m_1+\beta_1}(\mathbb{S}) \hookrightarrow h^{m_2+\beta_2}(\mathbb{S})$$

is dense and compact whenever $m_2 + \beta_2 < m_1 + \beta_1$. The small Hölder spaces can be described by the following intrinsic property:

Lemma 4.1.1 *Let $f \in C^{m+\beta}(\mathbb{S})$ be given. Then*

$$f \in h^{m+\beta}(\mathbb{S}) \quad \Leftrightarrow \quad \lim_{\tau \searrow 0} \sup_{0 < |t-s| < \tau} \frac{|f^{(m)}(t) - f^{(m)}(s)|}{|t-s|^\beta} = 0.$$

Proof The implication from left to right is immediate. The converse follows by using a partition for the unit circle $\mathbb{S}$. □

Assume that U is an open subset of $\mathbb{R}^2$. Recall that, for $k \in \mathbb{N} \cup \{\infty\}$ given, the set $BUC^k(U)$ denotes the space of all maps from U to $\mathbb{R}$ which have bounded and uniformly continuous derivatives up to order k. Given $u \in BUC^k(U)$, the derivative $\partial^\beta u$, $|\beta| \leq k$, can be extended to a uniformly continuous function on the closure $\overline{U}$.

Given $\beta \in (0,1)$, the space $BUC^{k+\beta}(U)$ consists of all functions $u \in BUC^k(U)$ having uniformly β−Hölder continuous derivatives of order k

$$\sum_{|\gamma|=k} \sup_{x \neq y} \frac{|\partial^\gamma u(x) - \partial^\gamma u(y)|}{|x-y|^\beta} < \infty.$$

Finally, we define $buc^{k+\beta}(U)$ to be the completion of $BUC^\infty(U)$ in $BUC^{k+\beta}(U)$.

4.2 Preparations and the well-posedness result

To handle the quasi–stationary problem (2.1) we parametrise first the boundary $\partial \Omega(t)$, $t \geq 0$. This will introduce, besides ψ and p, a third more important unknown: the function describing the boundary.

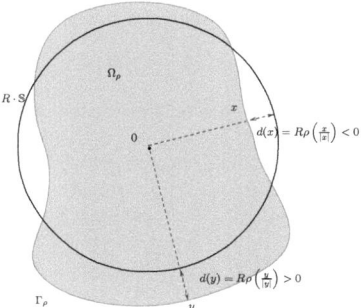

Figure 4.1: Parametrisation of the tumor domain

Let $\alpha \in (0,1)$ be fixed for the remainder of this work, and choose $R > 0$. We presuppose, in this chapter, that the initial state Ω_0 of the tumor is a perturbation of the disc $D(0, R)$. The set of admissible functions is given by

$$\mathcal{V} := \{\rho \in h^{4+\alpha}(\mathbb{S}) \,:\, \|\rho\|_{C(\mathbb{S})} < 1/4\}.$$

Obviously, $\mathcal{V}$ is an open neighbourhood of the zero function $0 \in h^{4+\alpha}(\mathbb{S})$. Given $\rho \in \mathcal{V}$, we define the $C^{4+\alpha}$ perturbation of the circle centred in 0 with radius R

$$\Gamma_\rho := \left\{ x \in \mathbb{R}^2 \,:\, |x| = R\left(1 + \rho\left(\frac{x}{|x|}\right)\right) \right\} = \{R(1+\rho(x))\,x \,:\, x \in \mathbb{S}\}.$$

The connected component of $\mathbb{R}^2$ which is bounded by the curve Γ_ρ is the set

$$\Omega_\rho := \left\{ x \in \mathbb{R}^2 \,:\, |x| < R\left(1 + \rho\left(\frac{x}{|x|}\right)\right) \right\},$$

with boundary $\partial \Omega_\rho = \Gamma_\rho$. Given $x \in \Gamma_\rho$, the real number $\rho(x/|x|)$ is the ratio of the signed distance from x to the circle $R \cdot \mathbb{S}$ and R (see Figure 4.1). It is suitable to represent Γ_ρ as the 0−level set of an appropriate function. For this, let $N_\rho : A(3R/4, 5R/4) \to \mathbb{R}$ be the function defined by

$$N_\rho(x) = |x| - R - R\rho(x/|x|), \quad x \in A(3R/4, 5R/4),$$

where $A(3R/4, 5R/4)$ is the annulus centred in 0 with radii $3R/4$ and $5R/4$

$$A(3R/4, 5R/4) := \{x \in \mathbb{R}^2 : 3R/4 < |x| < 5R/4\}.$$

Notice that $A(3R/4, 5R/4)$ is an open neighbourhood of Γ_ρ, and $\Gamma_\rho = N_\rho^{-1}(0)$. Let ν_ρ denote the outward normal at Γ_ρ. Since Γ_ρ is the 0−level set of N_ρ, the gradient ∇N_ρ and ν_ρ must be collinear vectors. Moreover, N_ρ is positive on the complement of $\overline{\Omega}_\rho$, hence $N_\rho(x + \lambda \nu_\rho(x)) > 0$ for all $x \in \Gamma_\rho$ and $\lambda > 0$. Derivating this relation with respect to λ at $\lambda = 0$ yields that $\nabla N_\rho \cdot \nu_\rho > 0$, hence

$$\nu_\rho = \frac{\nabla N_\rho}{|\nabla N_\rho|}.$$

To incorporate time let $T > 0$. Presuppose that the function $\rho \in C([0, T], \mathcal{V}) \cap C^1([0, T], h^{1+\alpha}(\mathbb{S}))$ describes the evolution of the tumor, which at time $t = 0$ is located at $\Omega(0) = \Omega_{\rho(0)}$. The normal velocity $V(t)$ of the moving boundary $\Gamma_{\rho(t)}$ is then given by the expression

$$V(t) = -\frac{\partial_t N_\rho}{|\nabla N_\rho|}.$$

This relation follows from the standard assumption that the interface moves along with the tumor and from relation $\Gamma_\rho = N_\rho^{-1}(0)$.

With these notations we come to the following moving boundary problem

$$\begin{cases} \Delta \psi &= f(\psi) & \text{in } \Omega_{\rho(t)}, \\ \Delta p &= 0 & \text{in } \Omega_{\rho(t)}, \\ \psi &= 1 & \text{on } \Gamma_{\rho(t)}, \\ p &= \kappa_{\Gamma_{\rho(t)}} - AG\frac{|x|^2}{4} & \text{on } \Gamma_{\rho(t)}, \\ \partial_t N_\rho &= -\left\langle G\nabla \psi - \nabla p - AG\frac{x}{2}, \nabla N_\rho \right\rangle & \text{on } \Gamma_{\rho(t)}, \\ \rho(0) &= \rho_0 & \text{on } \mathbb{S}, \end{cases} \quad (4.1)$$

for all $t \in [0, T]$.

A triple (ρ, ψ, p) is called a *classical Hölder solution* of (4.1) if

- $\rho \in C([0, T], \mathcal{V}) \cap C^1([0, T], h^{1+\alpha}(\mathbb{S}))$,

- $\psi(\cdot, t) \in buc^{2+\alpha}(\Omega_{\rho(t)})$ for all $0 \leq t \leq T$,

- $p(\cdot, t) \in buc^{2+\alpha}(\Omega_{\rho(t)})$ for all $0 \leq t \leq T$,

and (ρ, ψ, p) satisfies (4.1) pointwise.

Of major interest is to determine the mapping ρ which describes the evolution of the tumor. The functions ψ and p can be determined then as solutions of Dirichlet problems, cf. Lemma 4.3.1, Theorem 4.3.5, and Theorem 4.3.6. This is the reason why we shall also refer only to ρ as solution to (4.1). The first main result of the thesis, Theorem 4.2.1, is proved in Section 4.4.2 and guarantees local existence and uniqueness of classical solutions of problem (4.1).

Theorem 4.2.1 (Existence and uniqueness) *There exists an open neighbourhood $\mathcal{O}$ of 0 in $\mathcal{V}$ such that, for any initial data $\rho_0 \in \mathcal{O}$, there exists a maximal existence time $T := T(\rho_0) > 0$ and a unique classical solution ρ to problem (4.1) defined on $[0, T(\rho_0))$ which satisfies $\rho([0, T(\rho_0))) \subset \mathcal{O}$.*

4.3 The transformed problem

A fundamental difficulty in treating problem (4.1) is the fact that one has to work with unknown, variable domains Ω_ρ. We overcome this difficulty by transforming problem (4.1) on the unitary disc $\Omega := D(0, 1)$. Therefore, we define for all $\rho \in \mathcal{V}$ the mapping $\Theta_\rho : \mathbb{R}^2 \to \mathbb{R}^2$ by

$$\Theta_\rho(x) = \begin{cases} \dfrac{Rx}{|x|}\left(|x| + \varphi(|x| - 1)\rho\left(\dfrac{x}{|x|}\right)\right), & 0 < |x| < 2, \\ Rx, & \text{else}, \end{cases}$$

where $\varphi \in C^\infty(\mathbb{R}, [0, 1])$ satisfies

$$\varphi(r) = \begin{cases} 1, & |r| \leq 1/4, \\ 0, & |r| \geq 3/4, \end{cases}$$

and additionally $\max |\varphi'(r)| < 4$. In fact for $|x| \leq 1/4$ and for $|x| \geq 7/4$ we have $\Theta_\rho(x) = x$. Given $x \in \mathbb{S}$, the mapping $[0, \infty) \ni r \mapsto r + \varphi(r-1)\rho(x/|x|) \in [0, \infty)$ is strictly increasing and therefore bijective. The composition $\rho(\cdot/|\cdot|)$ has the same regularity properties as ρ on any subset of $\mathbb{R}^2$ which is bounded away from 0, and using the chain rule we have, cf. [29], that

$$\nabla\left(\rho\left(\frac{x}{|x|}\right)\right) = \rho'\left(\frac{x}{|x|}\right)\left(-\frac{x_2}{|x|^2}, \frac{x_1}{|x|^2}\right) \qquad (4.2)$$

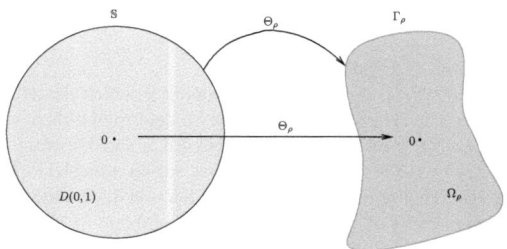

Figure 4.2: The Hanzawa diffeomorphism

for all $x \neq 0$. Consequently, Θ_ρ is a diffeomorphism, i.e. $\Theta_\rho \in \mathit{Diff}^{4+\alpha}(\Omega, \Omega_\rho) \cap \mathit{Diff}^{4+\alpha}(\mathbb{R}^2, \mathbb{R}^2)$. Such a diffeomorphism was first introduced by Hanzawa in [43] to study the Stefan problem, and it is therefore called *Hanzawa diffeomorphism*. Additionally, we have that $\Theta_\rho(\mathbb{S}) = \Gamma_\rho$ (see Figure 4.2). The push-forward operator induced by Θ_ρ is defined by

$$\Theta_\rho^* : BUC(\Omega_\rho) \to BUC(\Omega), \quad u \mapsto u \circ \Theta_\rho.$$

These operators allow us to transform the problem into an abstract Cauchy problem over $\mathbb{S}$. General results of the theory of maximal regularity, due to Sinestrari [53], can be used to prove existence of a unique classical solution, corresponding to small initial data. The solution to (4.1) is then obtained (see Lemma 4.3.1 below) using the pull-back operators defined by

$$\Theta_*^\rho : BUC(\Omega) \to BUC(\Omega_\rho), \quad v \mapsto v \circ \Psi_\rho,$$

where $\Psi_\rho := \Theta_\rho^{-1} = (f_\rho, g_\rho)$. The transformed operators $\mathcal{A}(\rho)$ and $\mathcal{B}$, are defined as follows. Given $\rho \in \mathcal{V}$, $\mathcal{A}(\rho) : buc^{2+\alpha}(\Omega) \to buc^\alpha(\Omega)$ is the differential operator given by

$$\mathcal{A}(\rho) := \Theta_\rho^* \circ \Delta \circ \Theta_*^\rho. \tag{4.3}$$

The trace operator $\mathcal{B} : \mathcal{V} \times buc^{2+\alpha}(\Omega) \times buc^{2+\alpha}(\Omega) \to h^{1+\alpha}(\mathbb{S})$ is defined by the following relation

$$\mathcal{B}(\rho, v, q) = \frac{1}{R} \operatorname{tr} \left\langle G \nabla(\Theta_*^\rho v)(\Theta_\rho) - \nabla(\Theta_*^\rho q)(\Theta_\rho) - \frac{AG}{2}\Theta_\rho, \nabla N_\rho(\Theta_\rho) \right\rangle, \tag{4.4}$$

with tr the *trace operator* on $\mathbb{S}$, i.e. $\operatorname{tr} v = v|_\mathbb{S}$ for $v \in BUC(\Omega)$. The curvature κ_{Γ_ρ} of Γ_ρ, $\rho \in \mathcal{V}$, is given by the relation

$$\kappa_{\Gamma_\rho}(\Theta_\rho) = \frac{(1+\rho)^2 + 2\rho'^2 - (1+\rho)\rho''}{R((1+\rho)^2 + \rho'^2)^{3/2}} =: \kappa(\rho).$$

It is not difficult to see that if (ρ, ψ, p) is a solution of (4.1) then $(\rho, v, q) = (\rho, \Theta_\rho^* \psi, \Theta_\rho^* p)$ solves pointwise the following transformed problem

$$\begin{cases} \mathcal{A}(\rho)v = f(v) & \text{in } \Omega, \\ v = 1 & \text{on } \mathbb{S}, \\ \mathcal{A}(\rho)q = 0 & \text{in } \Omega, \\ q = \kappa(\rho) - \dfrac{AGR^2}{4}(1+\rho)^2 & \text{on } \mathbb{S}, \\ \partial_t \rho = \mathcal{B}(\rho, v, q) & \text{on } \mathbb{S}, \\ \rho(0) = \rho_0. \end{cases} \quad (4.5)$$

The notion of solution for this problem is defined analogously to that of solution to (4.1). In fact the problems (4.1) and (4.5) are equivalent in the following sense:

Lemma 4.3.1 *Given $\rho_0 \in \mathcal{V}$ we have:*

(a) If (ρ, ψ, p) is a classical Hölder solution for (4.1), then $(\rho, \Theta_\rho^* \psi, \Theta_\rho^* p)$ is a classical Hölder solution for (4.5).

(b) If (ρ, v, q) is a classical Hölder solution for (4.5), then $(\rho, \Theta_*^\rho v, \Theta_*^\rho q)$ is a classical Hölder solution for (4.1).

Proof Given $\rho \in \mathcal{V}$ there exist positive constants K and δ depending only on ρ such that

$$\|\Theta_\rho - \Theta_{\tilde{\rho}}\|_{BUC^{4+\alpha}(\mathbb{R}^2)} \leq K \|\rho - \tilde{\rho}\|_{C^{4+\alpha}(\mathbb{S})} \quad (4.6)$$

for all $\tilde{\rho} \in \mathcal{V}$ with $\|\rho - \tilde{\rho}\|_{C^{4+\alpha}(\mathbb{S})} \leq \delta$. In fact we can choose K large enough such that the relation

$$\|\Theta_\rho^{-1} - \Theta_{\tilde{\rho}}^{-1}\|_{BUC^{4+\alpha}(\mathbb{R}^2)} \leq K \|\rho - \tilde{\rho}\|_{C^{4+\alpha}(\mathbb{S})} \quad (4.7)$$

holds for $\|\rho - \tilde{\rho}\|_{C^{4+\alpha}(\mathbb{S})} \leq \delta$. Indeed, given $\rho \in \mathcal{V}$, we compute, using the chain rule and the relation (4.2), that

$$f_{\rho,1}(\Theta_\rho(x)) = \begin{cases} \dfrac{1}{R(1+\varphi'\rho)} + \dfrac{x_2(-x_2\varphi\rho + x_2|x|\varphi'\rho + x_1\varphi\rho')}{R|x|^3\left(1 + \frac{1}{|x|}\varphi\rho\right)(1+\varphi'\rho)} & 0 < |x| < 2, \\ \dfrac{1}{R} & \text{else,} \end{cases}$$

$$f_{\rho,2}(\Theta_\rho(x)) = \begin{cases} -\dfrac{x_1(-x_2\varphi\rho + x_2|x|\varphi'\rho + x_1\varphi\rho')}{R|x|^3\left(1 + \frac{1}{|x|}\varphi\rho\right)(1+\varphi'\rho)} & 0 < |x| < 2, \\ 0 & \text{else,} \end{cases}$$

$$g_{\rho,1}(\Theta_\rho(x)) = \begin{cases} -\dfrac{x_2(-x_1\varphi\rho + x_1|x|\varphi'\rho - x_2\varphi\rho')}{R|x|^3\left(1 + \frac{1}{|x|}\varphi\rho\right)(1+\varphi'\rho)} & 0 < |x| < 2, \\ 0 & \text{else,} \end{cases}$$

$$g_{\rho,2}(\Theta_\rho(x)) = \begin{cases} \dfrac{1}{R(1+\varphi'\rho)} + \dfrac{x_1(-x_1\varphi\rho + x_1|x|\varphi'\rho - x_2\varphi\rho')}{R|x|^3\left(1 + \frac{1}{|x|}\varphi\rho\right)(1+\varphi'\rho)} & 0 < |x| < 2, \\ \dfrac{1}{R} & \text{else,} \end{cases}$$

where $\varphi = \varphi(|x| - 1)$ and $\rho = \rho(x/|x|)$. Relation (4.7) is now immediate. The assertion follows now due to relations (4.6) and (4.7), using arguments similar to the ones in Lemma 1.2 in [28]. □

Given $\rho \in \mathcal{V}$, the operator $\mathcal{A}(\rho)$, defined by (4.3), is linear and uniformly elliptic, with
$$\mathcal{A}(\rho)v = b_{ij}(\rho)v_{ij} + b_i(\rho)v_i, \quad \forall v \in buc^{2+\alpha}(\Omega),$$
where
$$b_{11}(\rho) = f_{\rho,1}^2(\Theta_\rho) + f_{\rho,2}^2(\Theta_\rho),$$
$$b_{12}(\rho) = f_{\rho,1}(\Theta_\rho)g_{\rho,1}(\Theta_\rho) + f_{\rho,2}(\Theta_\rho)g_{\rho,2}(\Theta_\rho),$$
$$b_{22}(\rho) = g_{\rho,1}^2(\Theta_\rho) + g_{\rho,2}^2(\Theta_\rho),$$
$$b_1(\rho) = f_{\rho,11}(\Theta_\rho) + f_{\rho,22}(\Theta_\rho),$$
$$b_2(\rho) = g_{\rho,11}(\Theta_\rho) + g_{\rho,22}(\Theta_\rho).$$

Indeed, let $\rho \in \mathcal{V}$ and $\xi = (\xi_1, \xi_2) \in \mathbb{R}^2$ with $|\xi|^2 = 1$. A simple calculation shows that

$$b_{ij}(\rho)\xi_i\xi_j = (f_{\rho,1}\xi_1 + g_{\rho,1}\xi_2)^2 + (f_{\rho,2}\xi_1 + g_{\rho,2}\xi_2)^2 = |\xi \cdot \partial\Psi_\rho(\Theta_\rho)|^2,$$

which yields the uniform ellipticity. Moreover, the ellipticity constants may be chosen uniformly for all ρ in some small ball in $h^{4+\alpha}(\mathbb{S})$. In addition, if g is the standard metric on $\mathbb{R}^2$, then the diffeomorphism Θ_ρ induces a Riemannian metric, $\Theta_\rho^* g$, on $\overline{\Omega}$, i.e.

$$\Theta_\rho^* g\big|_x(\xi,\zeta) := g\big|_{\Theta_\rho(x)}(T_x\Theta_\rho\xi, T_x\Theta_\rho\zeta) = \xi^T(\partial\Theta_\rho(x))^T \partial\Theta_\rho(x)\zeta, \qquad (4.8)$$

for $x \in \overline{\Omega}$ and tangent vectors $\xi, \zeta \in T_x\overline{\Omega}$. It can be checked that $\mathcal{A}(\rho)$ is exactly the *Laplace-Beltrami operator* corresponding to the Riemannian manifold $(\overline{\Omega}, \Theta_\rho^* g)$. Moreover, as the formulae above show, $\mathcal{A}$ depends analytically on ρ

$$\mathcal{A} \in C^\omega(\mathcal{V}, \mathcal{L}(buc^{2+\alpha}(\Omega), buc^\alpha(\Omega))). \qquad (4.9)$$

Lemma 4.3.2 *The mapping*

$$\mathcal{V} \ni \rho \mapsto \kappa(\rho) = \frac{(1+\rho)^2 + 2\rho'^2 - (1+\rho)\rho''}{R((1+\rho)^2 + \rho'^2)^{3/2}} \in h^{2+\alpha}(\mathbb{S})$$

is analytic. Moreover, $\partial\kappa(0)[\rho] = -(\rho + \rho'')/R$, *for all* $\rho \in h^{4+\alpha}(\mathbb{S})$.

Proof The analyticity is obvious. In order to compute the derivative one has only to calculate the gradient of a real valued function of three variables. □

We introduce now solution operators to some semilinear, respectively linear Dirichlet problems related to our transformed problem (4.5). In order to define the operators we discuss first the solvability of the following semilinear problem

$$\begin{cases} \Delta u = f(u) & \text{in } \Omega_\rho, \\ u = 1 & \text{on } \Gamma_\rho, \end{cases} \qquad (4.10)$$

with $\rho \in \mathcal{V}$. The ODE–technics presented in the previous chapter can no longer be applied since ρ may not be constant. We shall prove that (4.10) has solutions by using the well-known fixed point theorem due to Leray-Schauder (cf. [37, Theorem 11.3]).

Theorem 4.3.3 *Let T be a compact mapping of a Banach space $\mathbb{E}$ into itself, and suppose that there exists a constant M such that*

$$\|x\|_E \leq M$$

for all $x \in \mathbb{E}$ and $\sigma \in [0,1]$ satisfying $x = \sigma T x$. Then T has a fixed point.

In order to apply Theorem 4.3.3 to the Dirichlet problem (4.10), we fix $\rho \in \mathcal{V}$ and take the Banach space $\mathbb{E}$ to be the Hölder space $BUC^\alpha(\Omega_\rho)$. Furthermore, we consider the following continuously differentiable extension of f to the entire real line

$$g(t) = \begin{cases} f(t), & t \geq 0, \\ f'(0) \cdot t, & t \leq 0. \end{cases} \quad (4.11)$$

Notice that the derivative g' is positive and g is negative on the negative half axis. For all $u \in BUC^\alpha(\Omega_\rho)$, the operator $\mathfrak{T}$ is defined by letting $v = \mathfrak{T}u$ be the unique solution in $BUC^\alpha(\Omega_\rho)$ of the linear Dirichlet problem

$$\begin{cases} \Delta v = g(u) & \text{in } \Omega_\rho, \\ v = 1 & \text{on } \Gamma_\rho. \end{cases} \quad (4.12)$$

In fact, the solution belongs to $BUC^{2+\alpha}(\Omega_\rho)$ which is compactly embedded in $BUC^\alpha(\Omega_\rho)$. Consequently, $\mathfrak{T}$ is a compact operator. The solvability of the Dirichlet problem (4.10), in the space $BUC^{2+\alpha}(\Omega_\rho)$ is thus equivalent to the solvability of the equation $u = \mathfrak{T}u$ in the Banach space $\mathbb{E} = BUC^\alpha(\Omega_\rho)$.

The equation $u = \sigma \mathfrak{T}u$ in $\mathbb{E}$ is equivalent to the Dirichlet problem

$$\begin{cases} \Delta u = \sigma g(u) & \text{in } \Omega_\rho, \\ u = \sigma & \text{on } \Gamma_\rho \end{cases} \quad (4.13)$$

for all $\sigma \in [0,1]$. In order to apply the Leray-Schauder Theorem, we need to prove a priori estimates for the solutions to (4.13). In our context we need to prove just C^α-estimates for the solution of (4.13), which constitutes a great advantage compared with the theory in [37]. The methods presented in [37] apply very well in here. In fact, we prove even more:

Lemma 4.3.4 *There exists a positive constant M such that every solution $u \in BUC^{2+\alpha}(\Omega_\rho)$ of (4.13), with $\sigma \in [0,1]$, satisfies*

$$\|u\|_{BUC^1(\Omega_\rho)} \leq M, \quad (4.14)$$

with M independent of u and σ.

Proof For simplicity, we divide the proof in 3 steps.

Step 1. *Given* $u \in BUC^{2+\alpha}(\Omega_\rho)$ *a solution of* (4.13), *we have that*
$$0 \leq u \leq 1. \tag{4.15}$$
If $\sigma = 0$, it follows that $u = 0$. Let now $\sigma \in (0, 1]$. Assume by contradiction that there exists a $x_0 \in \Omega_\rho$ such that $u(x_0) = \min_{\Omega_\rho} u < 0$. Then, in view of $\Delta u(x_0) \geq 0$ and $g(u(x_0)) < 0$ we obtain a contradiction. Thus, $u \geq 0$. Therefore,
$$\begin{cases} \Delta u \geq 0 & \text{in } \Omega_\rho, \\ u = \sigma & \text{on } \partial\Omega_\rho. \end{cases}$$
The estimate (4.15) follows now from elliptic maximum principles cf. [37, Corollary 3.2].

Step 2. *There exists a positive constant* M_1 *with the property that every solution* $u \in BUC^{2+\alpha}(\Omega_\rho)$ *of* (4.13) *fulfills*
$$\|\nabla u\|_{C(\partial\Omega_\rho)} \leq M_1, \tag{4.16}$$
independently of $\sigma \in [0, 1]$.

Being a $C^{4+\alpha}$–domain, Ω_ρ satisfies an uniform exterior sphere condition with radius $r > 0$. Choose now $p_0 \in \partial\Omega_\rho$ and $q_0 = (q_{0_1}, q_{0_2}) \notin \overline{\Omega_\rho}$ such that $\overline{D}(q_0, r) \cap \overline{\Omega_\rho} = \{p_0\}$. For $a \in (0, 1)$ we consider the set
$$\mathcal{N} := \{p \in \Omega_\rho : d(p, D(q_0, r)) < a\},$$
and the mapping $d : \mathcal{N} \to \mathbb{R}_+$,
$$d(p) := d(p, D(q_0, r)) = \sqrt{(p_1 - q_{0_1})^2 + (p_2 - q_{0_2})^2} - r, \ \forall p = (p_1, p_1) \in \mathcal{N},$$
which is the distance from p to $D(q_0, r)$. The function d is smooth, i.e. $d \in BUC^\infty(\mathcal{N})$ and
$$d_i(p) = \frac{p_i - q_{0_i}}{|p - q_0|},$$
$$d_{ij}(p) = |p - q_0|^{-3}(|p - q_0|^2 \delta_{ij} - (p_i - q_{0_i})(p_j - q_{0_j})), \ \forall p \in \mathcal{N}.$$
Let $\overline{\nu}, k$ be two positive constants, to be fixed later, and $\varphi : \mathbb{R}_+ \to \mathbb{R}$ the function defined by
$$\varphi(d) = \frac{1}{\overline{\nu}} \log(1 + kd) \quad \text{for } d \geq 0.$$

The function φ is increasing and
$$\varphi'(d) = \frac{1}{\bar{\nu}} \cdot \frac{k}{1+kd}, \quad \varphi''(d) = -\frac{1}{\bar{\nu}} \cdot \frac{k^2}{(1+kd)^2} = -\bar{\nu}(\varphi'(d))^2.$$

We consider now the composition $w := \varphi \circ d \in BUC^\infty(\mathcal{N})$ and set $z := u - \sigma$. Obviously, z is the solution of the equation $\Delta z = \sigma g(u)$ in $\mathcal{N}$ and satisfies homogeneous Dirichlet boundary conditions on Γ_ρ. Since $w_i = \varphi' d_i$ and $w_{ij} = \varphi'' d_i d_j + \varphi' d_{ij}$ we obtain that

$$\Delta w \pm \sigma g(u) \leq \varphi''(d_1^2 + d_2^2) + \varphi'(d_{11} + d_{22}) + g(1) = -\bar{\nu}\varphi'^2 + \frac{1}{R}\varphi' + g(1)$$

$$\leq \frac{k}{R\bar{\nu}} - \frac{1}{\bar{\nu}}\frac{k^2}{(1+ka)^2} + g(1) \quad \text{in} \quad \mathcal{N}.$$

We define now the operator $\mathfrak{D} : BUC^{2+\alpha}(\mathcal{N}) \to BUC^\alpha(\mathcal{N})$ by

$$\mathfrak{D}w = \Delta w - \sigma g(u),$$

for $w \in BUC^{2+\alpha}(\mathcal{N})$. The constants a, k and $\bar{\nu}$ are to be chosen such that

$$\begin{cases} \mathfrak{D}w \leq 0 & \text{in} \quad \mathcal{N}, \\ w \geq z & \text{on} \quad \partial\mathcal{N}. \end{cases}$$

If $p \in \partial\mathcal{N}$ is a point with $d(p, D(q_0, r)) = a$, then we have

$$w(p) = \varphi(a) = \frac{1}{\bar{\nu}}\log(1+ka) > 1 \geq \max_{\Omega_\rho}|z| \Leftrightarrow ka \geq e^{\bar{\nu}} - 1.$$

Choosing $a = 1/k$ and $\bar{\nu} \in (0,1)$ such that $2 \geq e^{\bar{\nu}}$, then $w \geq z$ on the boundary portion $\{p \in \overline{\mathcal{N}} : d(p, D(q_0, r)) = a\}$. On the other boundary components of $\mathcal{N}$ the inequality holds, because $z = 0$ there and $w \geq 0$. Since $a = 1/k$, the upper bound of $\mathfrak{D}w$ found above satisfies

$$\frac{k}{R\bar{\nu}} - \frac{1}{\bar{\nu}}\frac{k^2}{(1+ka)^2} + g(1) = k\left(\frac{1}{R\bar{\nu}} - \frac{k}{4\bar{\nu}}\right) + g(1) \xrightarrow[k\to\infty]{} -\infty,$$

so we may fix $k > 1$ such that $\mathfrak{D}w < -1$.

Summarising, we found that the function $z - w \in BUC^{2+\alpha}(\mathcal{N})$ satisfies

$$\begin{cases} \Delta(z-w) = \mathfrak{D}z - \mathfrak{D}w \geq 0 & \text{in} \quad \mathcal{N}, \\ z - w \leq 0 & \text{on} \quad \partial\mathcal{N}. \end{cases}$$

Hence, $z \leq w$ in $\mathcal{N}$. With a, k and $\bar{\nu}$ as above we also have that

$$\begin{cases} \mathfrak{D}(-w) > 0 & \text{in } \mathcal{N}, \\ w > -z & \text{on } \partial\mathcal{N}, \end{cases}$$

and so $z + w \geq 0$ in $\mathcal{N}$. Consequently, we obtain that

$$w(p_0) = z(p_0) = 0 \quad \text{and} \quad -w \leq z \leq w \quad \text{in } \mathcal{N},$$

and, since $z = 0$ on Γ_ρ, we have the following estimate for the gradient of z in p_0

$$|\nabla z|(p_0) = |\langle \nabla z | \nu \rangle|(p_0) \leq |\langle \nabla w | \nu \rangle|(p_0).$$

The constants a, k and $\bar{\nu}$ do not depend on the boundary point p_0. Hence, in view of $|\nabla w|(p_0) = \varphi'(0)$, we get

$$|\nabla z| \leq \frac{k}{\bar{\nu}} \quad \text{on} \quad \Gamma_\rho,$$

which yields to the desired estimate.

Step 3. With M_1 the constant in (4.16) we have that every solution u of (4.13) satisfies the estimate

$$\|\nabla u\|_{BUC(\Omega_\rho)} \leq M_1. \tag{4.17}$$

Suppose there exists a point $p \in \Omega_\rho$ with the property that

$$\partial_k u(p) = \max_{\Omega_\rho} \partial_k u := K > \max_{\Gamma_\rho} \partial_k u,$$

where $k \in \{1, 2\}$. Let $w := \partial_k u$ and $r > 0$ such that $\overline{D}(p, r) \subset \Omega_\rho$. Further on, set $D := D(p, r)$. Obviously $w \in W^{1,2}(D)$. Let $\eta \in C_0^1(D)$ be a continuously differentiable function with compact support in D and $(\eta_l)_{l \in \mathbb{N}} \subset C_0^\infty(D)$, such that $\eta_l \to \eta$ and $\partial_m \eta_l \to \partial_m \eta$ for $m = 1, 2$ uniformly in D. Taking into account that

$$\nabla(\eta_l \nabla u) = \eta_l \nabla(\nabla u) + \nabla \eta_l \cdot \nabla u,$$

one obtains

$$0 = \int_D \eta_l \Delta u \, dx + \int_D \nabla \eta_l \cdot \nabla u \, dx$$

$$= \int_D \eta_l \sigma g(u) \, dx + \int_D (\partial_1 \eta_l \cdot \partial_1 u + \partial_2 \eta_l \cdot \partial_2 u) \, dx.$$

Replacing η_l by $\partial_k \eta_l$, and using the partial integration, the above relation becomes

$$0 = \int_D -\eta_l \sigma g'(u) w \, dx - \int_D (\partial_1 \eta_l \cdot \partial_1 w + \partial_2 \eta_l \cdot \partial_2 w) \, dx.$$

Letting $l \to \infty$, we obtain that

$$\int_D [\eta \sigma g'(u) w + \partial_1 \eta \cdot \partial_1 w + \partial_2 \eta \cdot \partial_2 w] \, dx = 0,$$

hence $w \in W^{1,2}(D)$ is a weak solution of the linear elliptic equation

$$\partial_i (\delta_{ij} \partial_j w) - \sigma g'(u) w = 0.$$

From the strong maximum principle, cf. [37, Theorem 8.19], we conclude that the function w must be constant in D. Hence, our assumption was false, and (4.17) follows at once.

$\square$

From Lemmas 4.3.2 and 4.3.4, and Theorem 4.3.3 we obtain for each $\rho \in \mathcal{V}$ a solution $u \in BUC^{2+\alpha}(\Omega_\rho)$ of problem (4.10). Using the same arguments as in the proof of Theorem 3.1.7 we may prove the uniqueness of this solution. Consequently, we have:

Theorem 4.3.5 *Given $\rho \in \mathcal{V}$, there exists a unique solution $\mathcal{T}(\rho) \in buc^{2+\alpha}(\Omega)$ of the semilinear Dirichlet problem*

$$\begin{cases} \mathcal{A}(\rho) v = f(v) & \text{in } \Omega, \\ v = 1 & \text{on } \mathbb{S}. \end{cases} \quad (4.18)$$

It holds that $\mathcal{T} \in C^1(\mathcal{V}, buc^{2+\alpha}(\Omega))$. Moreover, $\mathcal{T}^{-1}(\{u \in buc^{2+\alpha}(\Omega) : u > 0\})$ is an open neighbourhood of 0 in $h^{4+\alpha}$ and the restriction

$$\left[\mathcal{T}^{-1}(\{u \in buc^{2+\alpha}(\Omega) : u > 0\}) \ni \rho \mapsto \mathcal{T}(\rho) \in buc^{2+\alpha}(\Omega) \right]$$

is smooth.

Proof Denote by $u \in BUC^{2+\alpha}(\Omega_\rho)$ the unique solution of (4.10). Then $v := \Theta_\rho^* u \in BUC^{2+\alpha}(\Omega)$ is the unique solution of (4.18). Notice that $v \in BUC^\infty(\Omega)$ if $\rho \in C^\infty(\mathbb{S})$.

Hence, if we show that T is smooth, then we obtain that the solution $v \in buc^{2+\alpha}(\Omega_\rho)$ for all $\rho \in \mathcal{V}$. In order to prove the smoothness of T we define the operator $\mathfrak{J} : \mathcal{V} \times BUC^{2+\alpha}(\Omega) \to BUC^\alpha(\Omega) \times C^{2+\alpha}(\mathbb{S})$ by setting

$$\mathfrak{J}(\rho, u) = (\mathcal{A}(\rho)u - g(u), \operatorname{tr} u - 1),$$

where g is the extension of f defined in (4.11). The problem (4.18) can be rewritten as $\mathfrak{J}(\rho, u) = 0$, and T is a parametrisation for the 0-level set of $\mathfrak{J}$. It is easily verified that $\mathfrak{J}$ is continuously differentiable. Moreover, a simple computation shows that the linearisation of $\mathfrak{J}$ with respect to the u is given by

$$\partial_u \mathfrak{J}(\rho_0, u_0)[u] = (\mathcal{A}(\rho_0)u - g'(u_0)u, \operatorname{tr} u) \quad \text{for } u \in BUC^{2+\alpha}(\Omega).$$

Let now $(h, \varphi) \in BUC^\alpha(\Omega) \times C^{2+\alpha}(\mathbb{S})$ be given. Since g' is strictly positive and locally Lipschitz continuous, elliptic theory guarantees the existence of a unique solution of the boundary value problem

$$\begin{cases} \mathcal{A}(\rho_0)u - g'(u_0)u = h & \text{in } \Omega, \\ u = \varphi & \text{on } \mathbb{S}. \end{cases}$$

This implies that $\partial_u \mathfrak{J}(\rho, u)$ is a topological isomorphisms from $BUC^{2+\alpha}(\Omega)$ onto $BUC^\alpha(\Omega) \times C^{2+\alpha}(\mathbb{S})$. Hence, the implicit function theorem yields

$$T \in C^1(\mathcal{V}, BUC^{2+\alpha}(\Omega)).$$

Particularly, T maps into $buc^{2+\alpha}(\Omega)$. With the notation used in Section 3, $T(0) = \psi_R$, where ψ_R is the unique solution of problem (3.8). We have shown in Theorem 3.1.15 that ψ_R is positive, whence $T(\rho) > 0$ for all $\rho \in T^{-1}(\{u \in buc^{2+\alpha}(\Omega) : u > 0\})$, which is an open neighbourhood of 0 in $h^{4+\alpha}(\mathbb{S})$. By applying the Implicit function theorem to the mapping $\mathfrak{J}_0 : T^{-1}(\{u \in BUC^{2+\alpha}(\Omega) : u > 0\}) \to BUC^\alpha(\Omega) \times C^{2+\alpha}(\mathbb{S})$ with

$$\mathfrak{J}_0(\rho, u) = (\mathcal{A}(\rho)u - f(u), \operatorname{tr} u - 1),$$

as we did above, we find that T is even smooth on the open subset $T^{-1}(\{u \in buc^{2+\alpha}(\Omega) : u > 0\})$ of $h^{4+\alpha}(\mathbb{S})$. This completes the proof. □

Notice that our analysis in Section 3 played an important role in showing that

$\mathcal{T}$ is smooth. It is not clear that the solution of (4.18) is positive for arbitrary, not constant $\rho \in \mathcal{V}$. We enhance that $\mathcal{T}^{-1}(\{u \in buc^{2+\alpha}(\Omega) : u > 0\})$ is an open neighbourhood of the zero function in $h^{4+\alpha}(\mathbb{S})$. To simplify our notation we replace $\mathcal{V}$ by $\mathcal{T}^{-1}(\{u \in buc^{2+\alpha}(\Omega) : u > 0\})$ in the following, meaning also that $\mathcal{T}(\rho)$ is a positive function for all ρ in $\mathcal{V}$.

Theorem 4.3.6 *Given $\rho \in \mathcal{V}$, there exists a unique solution $\mathcal{S}(\rho) \in buc^{2+\alpha}(\Omega)$ of the Dirichlet problem*

$$\begin{cases} \mathcal{A}(\rho)q = 0 & \text{in } \Omega, \\ q = \kappa(\rho) - \dfrac{AGR^2}{4}(1+\rho)^2 & \text{on } \mathbb{S}. \end{cases} \tag{4.19}$$

Moreover, the mapping $[\mathcal{V} \ni \rho \mapsto \mathcal{S}(\rho) \in buc^{2+\alpha}(\Omega)]$ is smooth.

Proof Given $\rho \in \mathcal{V}$, the mapping

$$(\mathcal{A}(\rho), \text{tr}) : BUC^{2+\alpha}(\Omega) \to BUC^{\alpha}(\Omega) \times C^{2+\alpha}(\mathbb{S})$$

is a topological isomorphism from $BUC^{2+\alpha}(\Omega)$ onto $BUC^{\alpha}(\Omega) \times C^{2+\alpha}(\mathbb{S})$. It is well-known that the function mapping a bijective bounded linear operator onto its inverse is analytical; it can be expressed by a Neumann expansion in the neighbourhood of some other linear isomorphism.

Hence, in view of Lemma 4.3.2 and equation (4.9) it follows that

$$\mathcal{S}(\rho) = (\mathcal{A}(\rho), \text{tr})^{-1}\left(0, \kappa(\rho) - \dfrac{AGR^2}{4}(1+\rho)^2\right)$$

is analytic. As in the previous theorem we obtain that $\mathcal{S}(\rho) \in buc^{2+\alpha}(\Omega)$ for all $\rho \in \mathcal{V}$.

$\square$

4.4 The nonlinear Cauchy problem

We use now the solution operators defined in Theorem 4.3.5 and Theorem 4.3.6 to transform the system (4.5) into an abstract Cauchy problem on the unit circle

$\mathbb{S}$. Replacing in the third equation of (4.5) v by $\mathcal{T}(\rho)$, the solution to (4.18), and q by $\mathcal{S}(\rho)$, the solution to (4.19), we obtain that

$$\partial_t \rho = \Phi(\rho), \qquad \rho(0) = \rho_0, \qquad (4.20)$$

where

$$\Phi(\,\cdot\,) := \mathcal{B}(\,\cdot\,, \mathcal{T}(\,\cdot\,), \mathcal{S}(\,\cdot\,)) \qquad (4.21)$$

is a nonlinear and nonlocal operator of third order which depends nonlinearly on ρ. In order to prove Theorem 4.2.1 we shall prove that Φ is a smooth mapping, and in Section 4.6 we show that $\partial\Phi(0)$ generates a strongly continuous analytic semigroup in $\mathcal{L}(h^{1+\alpha}(\mathbb{S}))$ with definition domain $h^{4+\alpha}(\mathbb{S})$, that is

$$-\partial\Phi(0) \in \mathcal{H}(h^{4+\alpha}(\mathbb{S}), h^{1+\alpha}(\mathbb{S})).$$

The main result of this section is the following regularity result:

Theorem 4.4.1 *The operator Φ is smooth, i.e. $\Phi \in C^\infty(\mathcal{V}, h^{1+\alpha}(\mathbb{S}))$. Its derivative, $\partial\Phi(0)$, writes as the sum $\partial\Phi(0) = A_1 + A_2$, where*

$$A_1\rho := \frac{1}{R^3}\partial_\nu((\Delta, \mathrm{tr})^{-1}(0, \rho'')) \quad \text{and} \quad A_2 \in \mathcal{L}(h^{2+\alpha}(\mathbb{S}), h^{1+\alpha}(\mathbb{S})). \qquad (4.22)$$

In order to prove this theorem, we have to study first the regularity properties of the operator $\mathcal{B}$, defined by (4.4). It is convenient to we write this operator as a sum

$$\mathcal{B}(\rho, v, q) = \frac{G}{R}\mathcal{B}_1(\rho, v) - \frac{1}{R}\mathcal{B}_1(\rho, q) - \mathcal{B}_2(\rho),$$

where $\mathcal{B}_1$ and $\mathcal{B}_2$ are the operators defined by the formulae

$$\mathcal{B}_1 : \mathcal{V} \times buc^{2+\alpha}(\Omega) \to h^{1+\alpha}(\mathbb{S}), \quad \mathcal{B}_1(\rho, v) = \mathrm{tr}\,\langle \nabla(\Theta_*^\rho v), \nabla N_\rho \rangle\,(\Theta_\rho),$$

$$\mathcal{B}_2 : \mathcal{V} \to h^{1+\alpha}(\mathbb{S}), \quad \mathcal{B}_2(\rho) = \frac{AG}{2R}\,\mathrm{tr}\,\langle \Theta_\rho, \nabla N_\rho(\Theta_\rho) \rangle.$$

Using the chain rule we get

$$\partial\Phi(0)[\rho] = -\frac{1}{R}\partial\mathcal{B}_1(0, \mathcal{S}(0))\,[\rho, \partial\mathcal{S}(0)[\rho]]$$
$$+ \frac{G}{R}\partial\mathcal{B}_1(0, \mathcal{T}(0))[\rho, \partial\mathcal{T}(0)[\rho]] - \partial\mathcal{B}_2(0)[\rho].$$

We shall see that the first term of this sum is the important one, since it is a third order operator. The last two terms are of lower order and play, as we shall see, no role when studying the well-posedness of the abstract evolution equation (4.20).

Further on, we study the regularity properties of the operators $\mathcal{B}_i$, $i \in \{1, 2\}$, and determine their Fréchet derivatives. Consider first the operator $\mathcal{B}_1$. As we have seen in Section 3, the function $\mathcal{T}(0)$ is radially symmetric, and $\mathcal{S}(0)$ is constant, hence it suffices to determine $\partial \mathcal{B}_1(0, v_0)$ for a radially symmetric function $v_0 \in buc^{2+\alpha}(\Omega)$. Using relation (4.2) we get that

$$\nabla N_\rho(\Theta_\rho(x)) = x - \frac{\rho'(x)}{1+\rho(x)}(-x_2, x_1), \quad x \in \mathbb{S} \qquad (4.23)$$

for all $f \in \mathcal{V}$. Given $\rho \in \mathcal{V}$ and $v \in buc^{2+\alpha}(\Omega)$, the chain rule yields $\nabla(\Theta_*^\rho v)(\Theta_\rho) = \nabla v \cdot \partial \Psi_\rho(\Theta_\rho)$, so that

$$\mathcal{B}_1(\rho, v) = \operatorname{tr}\left[f_{\rho,1}(\Theta_\rho)x_1 + f_{\rho,2}(\Theta_\rho)x_2 + \right.$$
$$\left. + \frac{\rho'}{1+\rho}(f_{\rho,1}(\Theta_\rho)x_2 - f_{\rho,2}(\Theta_\rho)x_1)\right]v_1$$
$$+ \operatorname{tr}\left[g_{\rho,1}(\Theta_\rho)x_1 + g_{\rho,2}(\Theta_\rho)x_2 + \right.$$
$$\left. + \frac{\rho'}{1+\rho}(g_{\rho,1}(\Theta_\rho)x_2 - g_{\rho,2}(\Theta_\rho)x_1)\right]v_2.$$

Furthermore, being on $\mathbb{S}$, the formulae for the partial derivatives of the components of the diffeomorphism Ψ_ρ, which we determined in Lemma 4.3.1, simplify to

$$\partial\Psi(\Theta_\rho(x)) = \begin{bmatrix} \dfrac{1+x_1^2\rho(x)+x_1x_2\rho'(x)}{R(1+\rho(x))} & \dfrac{x_1x_2\rho(x)-x_1^2\rho'(x)}{R(1+\rho(x))} \\ \dfrac{x_1x_2\rho(x)+x_2^2\rho'(x)}{R(1+\rho(x))} & \dfrac{1+x_2^2\rho(x)-x_1x_2\rho'(x)}{R(1+\rho(x))} \end{bmatrix}$$

for all $x \in \mathbb{S}$. We are now ready to state:

Lemma 4.4.2 *The nonlinear operator $\mathcal{B}_1$ is analytic, i.e.*

$$\mathcal{B}_1 \in C^\omega(\mathcal{V} \times buc^{2+\alpha}(\Omega), h^{1+\alpha}(\mathbb{S})).$$

Given $v_0 \in buc^{2+\alpha}(\Omega)$ a radially symmetric function, we have
$$\partial \mathcal{B}_1(0, v_0)[\rho, w] = \frac{1}{R} \operatorname{tr} \partial_\nu w \qquad (4.24)$$
for all $[\rho, w] \in h^{4+\alpha}(\mathbb{S}) \times buc^{2+\alpha}(\Omega)$, where $\nu = \nu_0$ is the unit exterior normal vector field at $\mathbb{S}$.

Proof The regularity assertion is an obvious consequence of the relations determined above. Let $[\rho, w] \in h^{4+\alpha}(\mathbb{S}) \times buc^{2+\alpha}(\Omega)$ be given. We prove that
$$\partial \mathcal{B}_1(0, v_0)[\rho, w] = \operatorname{tr} \langle \nabla \left(\Theta^0_* w \right), \nabla N_0 \rangle (\Theta_0). \qquad (4.25)$$
Indeed, we have that
$$\mathcal{B}_1(\rho, w + v_0) - \mathcal{B}_1(0, v_0) - \operatorname{tr} \langle \nabla \left(\Theta^0_* w \right), \nabla N_0 \rangle (\Theta_0) =$$
$$= \operatorname{tr} \langle \nabla (\Theta^\rho_* w), \nabla N_\rho \rangle (\Theta_\rho) + \operatorname{tr} \langle \nabla (\Theta^\rho_* v_0), \nabla N_\rho \rangle (\Theta_\rho)$$
$$- \operatorname{tr} \langle \nabla (\Theta^0_* v_0), \nabla N_0 \rangle (\Theta_0) - \operatorname{tr} \langle \nabla (\Theta^0_* w), \nabla N_0 \rangle (\Theta_0)$$
$$= \operatorname{tr} \langle \nabla (\Theta^\rho_* w), \nabla N_\rho \rangle (\Theta_\rho) - \operatorname{tr} \langle \nabla (\Theta^0_* w), \nabla N_0 \rangle (\Theta_0)$$
$$+ \operatorname{tr} \langle \nabla (\Theta^\rho_* v_0), \nabla N_\rho \rangle (\Theta_\rho) - \operatorname{tr} \langle \nabla (\Theta^0_* w), \nabla N_0 \rangle (\Theta_0).$$

We split the relation in two sums and estimate each of them separately. For the first sum we get
$$\langle \nabla (\Theta^\rho_* w), \nabla N_\rho \rangle (\Theta_\rho) - \langle \nabla (\Theta^0_* w), \nabla N_0 \rangle (\Theta_0) =$$
$$= w_1 \Big\{ [f_{\rho,1}(\Theta_\rho) - f_{0,1}(\Theta_0)] x_1 + [f_{\rho,2}(\Theta_\rho) - f_{0,2}(\Theta_0)] x_2$$
$$+ \frac{\rho'}{1+\rho} f_{\rho,1}(\Theta_\rho) x_2 - \frac{\rho'}{1+\rho} f_{\rho,2}(\Theta_\rho) x_1 \Big\}$$
$$+ w_2 \Big\{ [g_{\rho,1}(\Theta_\rho) - g_{0,1}(\Theta_0)] x_1 + [g_{\rho,2}(\Theta_\rho) - g_{0,2}(\Theta_0)] x_2$$
$$+ \frac{\rho'}{1+\rho} g_{\rho,1}(\Theta_\rho) x_2 - \frac{\rho'}{1+\rho} g_{\rho,2}(\Theta_\rho) x_1 \Big\}$$
$$= O(\|[\rho, w]\|^2_{C^{4+\alpha}(\mathbb{S}) \times C^{2+\alpha}(\Omega)}),$$

for small $[\rho, w] \in h^{4+\alpha}(\mathbb{S}) \times buc^{2+\alpha}(\Omega)$. In order to estimate the second term of the relation we shall use the property of v_0 to be radially symmetric. Thus, we obtain

$$\langle \nabla(\Theta_*^\rho v_0), \nabla N_\rho \rangle(\Theta_\rho) - \langle \nabla(\Theta_*^0 v_0), \nabla N_0 \rangle(\Theta_0) =$$

$$= v_{0,1} \left\{ [f_{\rho,1}(\Theta_\rho) - f_{0,1}(\Theta_0)]x_1 + [f_{\rho,2}(\Theta_\rho) - f_{0,2}(\Theta_0)]x_2 \right.$$

$$\left. + \frac{\rho'}{1+\rho} f_{\rho,1}(\Theta_\rho) x_2 - \frac{\rho'}{1+\rho} f_{\rho,2}(\Theta_\rho) x_1 \right\}$$

$$+ v_{0,2} \left\{ [g_{\rho,1}(\Theta_\rho) - g_{0,1}(\Theta_0)]x_1 + [g_{\rho,2}(\Theta_\rho) - g_{0,2}(\Theta_0)]x_2 \right.$$

$$\left. + \frac{\rho'}{1+\rho} g_{\rho,1}(\Theta_\rho) x_2 - \frac{\rho'}{1+\rho} g_{\rho,2}(\Theta_\rho) x_1 \right\}$$

$$= v_{0,1} \left\{ \frac{-x_2^2 \rho + x_1 x_2 \rho'}{R(1+\rho)} x_1 + \frac{x_1 x_2 \rho - x_1^2 \rho'}{R(1+\rho)} x_2 + \right.$$

$$\left. + \frac{\rho'}{1+\rho} \left[\frac{1 + x_1^2 \rho + x_1 x_2 \rho'}{R(1+\rho)} x_2 - \frac{x_1 x_2 \rho - x_1^2 \rho'}{R(1+\rho)} x_1 \right] \right\}$$

$$+ v_{0,2} \left\{ \frac{x_1 x_2 \rho + x_2^2 \rho'}{R(1+\rho)} x_1 - \frac{x_1^2 \rho + x_1 x_2 \rho'}{R(1+\rho)} x_2 + \right.$$

$$\left. + \frac{\rho'}{1+\rho} \left[\frac{x_1 x_2 \rho + x_2^2 \rho'}{R(1+\rho)} x_2 - \frac{1 + x_2^2 \rho - x_1 x_2 \rho'}{R(1+\rho)} x_1 \right] \right\} =$$

$$= v_{0,1} \frac{x_2 \rho' + x_1 \rho'^2}{R(1+\rho)^2} + v_{0,2} \frac{-x_1 \rho' + x_2 \rho'^2}{R(1+\rho)^2}$$

$$= (v_{0,1} x_1 + v_{0,2} x_2) \underbrace{\frac{\rho'^2}{R(1+\rho)^2}}_{=O(\|\rho\|_{C^{4+\alpha}(\mathbb{S})}^2)} + \underbrace{(v_{0,1} x_2 - v_{0,2} x_1)}_{=0} \frac{\rho'}{R(1+\rho)^2},$$

where we write $v_{0,i}$, $1 \leq i \leq 2$, for the partial derivatives of v_0, and we used the relation $v_{0,1}x_2 - v_{0,2}x_1 = 0$ on $\mathbb{S}$. This equality is obtained by differentiating the constant function $[\mathbb{R} \ni s \mapsto v_0(\cos(s), \sin(s))]$.

Thus, for small $[\rho, w] \in h^{4+\alpha}(\mathbb{S}) \times buc^{2+\alpha}(\Omega)$, we have the following estimate

$$||\mathcal{B}_1(\rho, w + v_0) - \mathcal{B}_1(0, v_0) - \text{tr}\langle \nabla (\Theta_*^0 w), \nabla N_0\rangle(\Theta_0)||_{C^{1+\alpha}(\mathbb{S})} =$$
$$= O(|||[\rho, w]||^2_{C^{4+\alpha}(\mathbb{S}) \times BUC^{2+\alpha}(\Omega)}),$$

which leads to (4.25). We infer from (4.23) that $\nabla N_0(\Theta_0) = \nu$. Since $\Psi_0 = R^{-1}\,\text{id}_{\mathbb{R}^2}$, we get that $\nabla(\Theta_*^0 w) = R^{-1}\nabla w$ and the proof is completed. $\square$

We consider now the operator $\mathcal{B}_2$. Since $\Theta_\rho(x)$ is a collinear vector with x and $(-x_2, x_1)$ is orthogonal on x, we obtain from relation (4.23) that

$$\mathcal{B}_2(\rho) = \frac{AG}{2R}\langle R(1+\rho)x, x\rangle = \frac{AG}{2}(1+\rho) \qquad (4.26)$$

for all $\rho \in \mathcal{V}$. Consequently, we have proved:

Lemma 4.4.3 *The operator $\mathcal{B}_2$ is analytic, i.e. $\mathcal{B}_2 \in C^\omega(\mathcal{V}, h^{1+\alpha}(\mathbb{S}))$. Moreover, we have*

$$\partial\mathcal{B}_2(0)[\rho] = \frac{AG}{2}\rho \qquad (4.27)$$

for all $\rho \in h^{4+\alpha}(\mathbb{S})$.

Proof The proof follows immediately from relation (4.26). $\square$

To finish the preparations for the proof of Theorem 4.4.1 one more step must be done. We have to determine the Fréchet derivative in 0 of the analytic solution operator defined in Theorem 4.3.6.

Lemma 4.4.4 *Given $\rho \in \mathcal{V}$, the map $\partial\mathcal{S}(0)[\rho] \in buc^{2+\alpha}(\Omega)$ is the unique solution of the linear Dirichlet problem*

$$\begin{cases} \Delta z = 0 & \text{in } \Omega, \\ z = -\left(\frac{1}{R} + \frac{AGR^2}{2}\right)\rho - \frac{1}{R}\rho'' & \text{on } \mathbb{S}. \end{cases} \qquad (4.28)$$

Proof Given $\rho \in h^{4+\alpha}(\mathbb{S})$, we let $z = z(\rho)$ denote the solution of (4.28). Our goal is to prove the following estimate

$$\|\mathcal{S}(\rho) - \mathcal{S}(0) - z\|_{BUC^{2+\alpha}(\Omega)} = O(\|\rho\|^2_{C^{4+\alpha}(\mathbb{S})}) \qquad (4.29)$$

for ρ sufficiently small in $h^{4+\alpha}(\mathbb{S})$. Recall that $\mathcal{S}(0) = 1/R + AGR^2/4$. Furthermore, since $\Psi_0 = R^{-1}\operatorname{id}_{\mathbb{R}^2}$, we have that $\mathcal{A}(0) = R^{-2}\Delta$. Setting $w := \mathcal{S}(\rho) - \mathcal{S}(0) - z$, we obtain, in view of $\mathcal{A}(\rho)\mathcal{S}(\rho) - \mathcal{A}(0)\mathcal{S}(0) - \mathcal{A}(0)z = 0$, the following equation in Ω

$$b_{ij}(\rho)w_{ij} + (b_{ij}(\rho) - b_{ij}(0))z_{ij} + b_i(\rho)w_i + (b_i(\rho) - b_i(0))z_i = 0,$$

where $b_{ij}(\rho)$ and $b_i(\rho)$, $1 \leq i,j \leq 2$, are the coefficients of $\mathcal{A}(\rho)$. On the boundary $\mathbb{S}$ of Ω we have, in view of Lemma 4.3.2, that

$$\begin{aligned}\mathcal{S}(\rho) - \mathcal{S}(0) - z &= \kappa(\rho) - \frac{AGR^2}{4}(1+\rho)^2 - \kappa(0) + \frac{AGR^2}{4} \\ &\quad - \partial\kappa(0)[\rho] + \frac{AGR^2}{2}\rho \\ &= \kappa(\rho) - \kappa(0) - \partial\kappa(0)[\rho] - \frac{AGR^2}{4}\rho^2 = O(\|\rho\|^2_{C^{4+\alpha}(\mathbb{S})}),\end{aligned}$$

for small $\rho \in h^{4+\alpha}(\mathbb{S})$. Therefore, w is the solution of the following problem

$$\begin{cases} \mathcal{A}(\rho)w = \underbrace{-z_{ij}(b_{ij}(\rho) - b_{ij}(0)) - z_i(b_i(\rho) - b_i(0))}_{=:\, f} \\ w = \varphi \end{cases}$$

and there exists $\Lambda > 0$ such that the estimates

- $\|b_{ij}\|_{BUC^\alpha(\Omega)} < \Lambda,\quad \|b_i\|_{BUC^\alpha(\Omega)} < \Lambda,$
- $\|f\|_{BUC^\alpha(\Omega)} < \Lambda\|\rho\|^2_{C^{4+\alpha}(\mathbb{S})},$
- $\|\varphi\|_{C^{2+\alpha}(\mathbb{S})} < \Lambda\|\rho\|^2_{C^{4+\alpha}(\mathbb{S})},$

hold, provided ρ is sufficiently small. As we mentioned before, the operator $\mathcal{A}(\rho)$ is uniformly elliptic for all ρ in some small neighbourhood of 0 in $h^{4+\alpha}(\mathbb{S})$, with uniform ellipticity constants in ρ. Using estimates for the solutions of elliptic problems (see [37], estimates (3.12) and (6.36)) we obtain

$$\|w\|_{BUC^{2+\alpha}(\Omega)} < c(\|f\|_{BUC^\alpha(\Omega)} + \|\varphi\|_{C^{2+\alpha}(\mathbb{S})}) \leq C\|\rho\|^2_{C^{4+\alpha}(\mathbb{S})},$$

which shows that $z = \partial \mathcal{S}(0)[\rho]$. □

We prove now the main result of this section:

Proof *(Proof of Theorem 4.4.1)* The regularity assumption follows directly from Theorem 4.3.5, Theorem 4.3.6, Lemma 4.4.2, and Lemma 4.4.3. Moreover, combining these results we have that

$$\partial \Phi(0)[\rho] = \frac{1}{R^3} \partial_\nu \left((\Delta, \mathrm{tr})^{-1}(0, \rho'') \right)$$
$$+ \left(\frac{1}{R^3} + \frac{AG}{2} \right) \partial_\nu \left((\Delta, \mathrm{tr})^{-1}(0, \rho) \right) + \frac{G}{R^2} \partial_\nu (\partial \mathcal{T}(0)[\rho]) - \frac{AG}{2} \rho.$$

The proof of Theorem 4.3.5 shows in fact that, the operator $\mathcal{T}$ is well-defined as operator $\mathcal{T} : \{\rho \in h^{2+\alpha}(\mathbb{S}) : \|\rho\|_{C(\mathbb{S})} < 1/4\} \to buc^{2+\alpha}(\Omega)$. Whence, the operator defined by

$$A_2 \rho := \left(\frac{1}{R^3} + \frac{AG}{2} \right) \partial_\nu \left((\Delta, \mathrm{tr})^{-1}(0, \rho) \right) + \frac{G}{R^2} \partial_\nu (\partial \mathcal{T}(0)[\rho]) - \frac{AG}{2} \rho$$

for $\rho \in h^{2+\alpha}(\mathbb{S})$, belongs to $\mathcal{L}(h^{2+\alpha}(\mathbb{S}), h^{1+\alpha}(\mathbb{S}))$, and we are done. □

4.5 Besov spaces and Fourier multiplier operators

In this subsection we are interested to characterise Fourier multiplier operators between the Hölder spaces we deal with. We use the standard notation $\mathcal{D}(\mathbb{S}, \mathbb{C})$ when we refer to the space consisting of complex-valued infinitely differentiable functions $\mathbb{S}$, i.e. $\mathcal{D}(\mathbb{S}, \mathbb{C}) = C^\infty(\mathbb{S}, \mathbb{C})$. Given $n \in \mathbb{N}$, the mapping

$$\|\varphi\|_n := \max_{k \leq n} \max_{\mathbb{S}} |\varphi^{(k)}|$$

defines a norm on $\mathcal{D}(\mathbb{S}, \mathbb{C})$, and endowed with the metric

$$d(\varphi, \psi) := \sum_{n=0}^{\infty} \frac{1}{2^n} \frac{\|\varphi - \psi\|_n}{1 + \|\varphi - \psi\|_n}, \quad \varphi, \psi \in \mathcal{D}(\mathbb{S}, \mathbb{C}),$$

$\mathcal{D}(\mathbb{S}, C)$ is a Fréchet space (see [46, Exercise 7.22]). This is in contrast to the situation when we consider $\mathcal{D}(\mathcal{O})$ for some open set $\mathcal{O} \subset \mathbb{R}^n$. It is known that $\mathcal{D}(\mathcal{O})$ is not metrizable in this case cf. [46, Remark 6.9].

Let $\mathcal{D}'(\mathbb{S}, \mathbb{C})$ denote the topological dual of $\mathcal{D}(\mathbb{S}, \mathbb{C})$. Given $f \in \mathcal{D}'(\mathbb{S}, \mathbb{C})$, the complex numbers
$$\widehat{f}(k) = f(e^{-ikt}), \ k \in \mathbb{Z}$$
are the *Fourier coefficients* of the linear functional f. For any $f \in \mathcal{D}'(\mathbb{S}, \mathbb{C})$ there exist constants $c_f > 0$ and $N_f \in \mathbb{N}$ such that
$$|\widehat{f}(k)| \leq c_f (1 + |k|)^{N_f} \quad \text{for all} \quad k \in \mathbb{Z}. \tag{4.30}$$

Indeed, since f is continuous and the family of norms $\{\|\cdot\|_n\}_n$ is ascending we find constants $\widetilde{c}_f$ and $\widetilde{N}_f$ such that
$$|f(\varphi)| \leq \widetilde{c}_f \|\varphi\|_{\widetilde{N}_f}$$
for all $\varphi \in \mathcal{D}(\mathbb{S}, \mathbb{C})$. It is suitable to introduce now Sobolev spaces over $\mathbb{S}$. Given $r \geq 0$, the *Sobolev space* $H^r(\mathbb{S}, \mathbb{C})$ is defined by
$$H^r(\mathbb{S}, \mathbb{C}) := \left\{ f \in L^2(\mathbb{S}, \mathbb{C}) : \sum_{k \in \mathbb{Z}} (1 + |k|^2)^r |\widehat{f}(k)|^2 < \infty \right\}.$$

Endowed with the scalar product
$$\langle f, g \rangle := \sum_{k \in \mathbb{Z}} (1 + |k|^2)^r \widehat{f}(k) \overline{\widehat{g}(k)}$$
it is a Hilbert space. If $r > 1/2$, then $H^{n+r}(\mathbb{S}, \mathbb{C}) \hookrightarrow C^n(\mathbb{S}, \mathbb{C})$, hence
$$H^{\widetilde{N}_f + 1}(\mathbb{S}, \mathbb{C}) \hookrightarrow C^{\widetilde{N}_f}(\mathbb{S}, \mathbb{C}).$$

Setting $N_f := \widetilde{N}_f + 1$, we get (4.30).

We prove now that the series $\sum_{k \in \mathbb{Z}} \widehat{f}(k) e^{ikt}$ converges to f in the weak*-topology of $\mathcal{D}'(\mathbb{S}, \mathbb{C})$. Given $\varphi \in \mathcal{D}(\mathbb{S}, \mathbb{C})$, we obtain in view of (4.30) that
$$\sum_{k \in \mathbb{Z}} \widehat{f}(k) e^{ikt}[\varphi] = \sum_{k \in \mathbb{Z}} \widehat{f}(k) \int_{\mathbb{S}} \varphi(t) e^{ikt} \, dt = \sum_{k \in \mathbb{Z}} \widehat{f}(k) \widehat{\varphi}(-k).$$

Moreover, for $\varphi \in \mathcal{D}(\mathbb{S}, \mathbb{C})$, the Fourier series $\sum_{k \in \mathbb{Z}} \widehat{\varphi}(k) e^{ikt}$ converges to φ in $C^n(\mathbb{S}, \mathbb{C})$ for all $n \in \mathbb{N}$, thus also in $\mathcal{D}(\mathbb{S}, \mathbb{C})$. Consequently,

$$f(\varphi) = f\left(\sum_{k \in \mathbb{Z}} \widehat{\varphi}(k) e^{ikt}\right) = \sum_{k \in \mathbb{Z}} \widehat{\varphi}(k) \widehat{f}(-k)$$

and the assertion is now immediate. Conversely, if $(a_k)_{k \in \mathbb{Z}}$ satisfies (4.30) for some $N \in \mathbb{N}$, then the series $\sum_{k \in \mathbb{Z}} a_k e^{ikt}$ converges in $\mathcal{D}'(\mathbb{S}, \mathbb{C})$ to a distribution on $\mathcal{D}(\mathbb{S}, \mathbb{C})$.

Let $(\phi_j)_{j \geq 0}$ be a sequence in the Schwartz space $\mathcal{S}(\mathbb{R})$ with the following properties:

(i) $\operatorname{supp} \phi_0 \subset [-2, 2]$, $\quad \operatorname{supp} \phi_j \subset \{t : 2^{j-1} \leq |t| \leq 2^{j+1}\}$, $j \geq 1$;

(ii) $\displaystyle\sum_{j \in \mathbb{N}} \phi_j = 1$, on $\mathbb{R}$;

(iii) $\forall k \in \mathbb{N} \, \exists \, c_k > 0 : \quad 2^{kj} \|\phi_j^{(k)}\|_0 \leq c_k, \, \forall j \in \mathbb{N}$.

For $s > 0$ we define

$$B_{\infty,\infty}^s(\mathbb{S}) := \left\{ f = \sum_{k \in \mathbb{Z}} \widehat{f}(k) e^{ikt} \in \mathcal{D}'(\mathbb{S}, \mathbb{C}) : \right.$$

$$\left. \|f\|_{B_{\infty,\infty}^s(\mathbb{S})}^{(\phi_j)} := \sup_{j \in \mathbb{N}} 2^{sj} \left\| \sum_{k \in \mathbb{Z}} \phi_j(k) \widehat{f}(k) e^{ikt} \right\|_{C(\mathbb{S})} < \infty \right\},$$

the so called *Besov space*.

It can be shown, cf. [50] that $B_{\infty,\infty}^s(\mathbb{S})$ is a Banach space. Moreover, the norm $\|\cdot\|_{B_{\infty,\infty}^s(\mathbb{S})}^{(\phi_j)}$ depends on the sequence (ϕ_j) in the following sense. If $(\psi_j)_{j \geq 0} \subset \mathcal{S}(\mathbb{R})$ satisfies the conditions $(i)-(iii)$ mentioned above, then the norm $\|\cdot\|_{B_{\infty,\infty}^s(\mathbb{S})}^{(\psi_j)}$ is equivalent to $\|\cdot\|_{B_{\infty,\infty}^s(\mathbb{S})}^{(\phi_j)}$.

One can also see, cf. [50, Theorem 3.5.4 (i)] that, if $s > 0$ is not an integer, then $B_{\infty,\infty}^s(\mathbb{S}) = C^s(\mathbb{S}, \mathbb{C})$. This description of the Hölder spaces using dyadic decomposition is very useful because it simplifies things when one tries to characterise Fourier multiplier operators between Hölder spaces.

The main result of this section is the following theorem which appeared first in [9] and was used to describe operator valued Fourier multipliers on periodic Besov spaces, but in the special case when the Besov spaces have the same orders.

It was generalised in [29] for Fourier multiplier operators between complex valued periodic Besov spaces of different order and used to prove that the linearisation of some nonlinear and nonlocal operator generates a strongly continuous and analytic semigroup.

Theorem 4.5.1 *Let r, s be two positive constants and let $(M_k)_{k \in \mathbb{Z}} \subset \mathbb{C}$ be a sequence satisfying the following conditions*

$$(i) \quad s_1 := \sup_{k \in \mathbb{Z} \setminus \{0\}} |k|^{r-s} |M_k| < \infty,$$

$$(ii) \quad s_2 := \sup_{k \in \mathbb{Z} \setminus \{0\}} |k|^{r-s+1} |M_{k+1} - M_k| < \infty,$$

$$(iii) \quad s_3 := \sup_{k \in \mathbb{Z} \setminus \{0\}} |k|^{r-s+2} |M_{k+2} - 2M_{k+1} + M_k| < \infty.$$

The mapping

$$\sum_{k \in \mathbb{Z}} \widehat{h}(k) e^{ikt} \stackrel{\mathcal{M}}{\mapsto} \sum_{k \in \mathbb{Z}} M_k \widehat{h}(k) e^{ikt},$$

belongs then to $\mathcal{L}(B^s_{\infty,\infty}(\mathbb{S}), B^r_{\infty,\infty}(\mathbb{S}))$.

Before starting the proof of Theorem 4.5.1, we prove first an useful lemma. It is well known (cf. [6, Corollary 9.13]) that if $f \in C(\mathbb{R}, \mathbb{C})$ has compact support and $\mathcal{F}^{-1} f \in L^1(\mathbb{R}, \mathbb{C})$, then $f = \mathcal{F} \mathcal{F}^{-1} f$. We have denoted by $\mathcal{F}$ and $\mathcal{F}^{-1}$ the Fourier transform and the inverse Fourier transform operator, respectively, i.e.

$$\mathcal{F} f(\xi) = (2\pi)^{-1/2} \int_{\mathbb{R}} f(t) e^{-it\xi} \, dx,$$

and $\mathcal{F}^{-1} f(\xi) = \mathcal{F} f(-\xi)$ for $f \in L^1(\mathbb{R}, \mathbb{C})$ and $\xi \in \mathbb{R}$.

Lemma 4.5.2 *Let $M \in C(\mathbb{R})$ be a function with compact support with the property that $\mathcal{F}^{-1} M \in L^1(\mathbb{R}, \mathbb{C})$. Given a trigonometric polynomial*

$$h = \sum_{k \in \mathbb{Z}} \widehat{h}(k) e^{ikt}$$

we have that

$$\left\|\sum_{k\in\mathbb{Z}} M(k)\widehat{h}(k)e^{ikt}\right\|_{C(\mathbb{S},\mathbb{C})} \leq (2\pi)^{-1/2}\|\mathcal{F}^{-1}M\|_{L^1}\|h\|_{C(\mathbb{S},\mathbb{C})}. \tag{4.31}$$

Proof Since h is a trigonometric polynomial we have that $\widehat{h}(k) = 0$ for k large enough. We are therefore allowed to commute the integral and sum symbol to obtain

$$\sum_{k\in\mathbb{Z}} M(k)\widehat{h}(k)e^{ikt} = \sum_{k\in\mathbb{Z}} \mathcal{F}(\mathcal{F}^{-1}M)(k)\widehat{h}(k)e^{ikt}$$

$$= (2\pi)^{-1/2}\sum_{k\in\mathbb{Z}}\int_{\mathbb{R}} \mathcal{F}^{-1}M(s)e^{ik(t-s)}\,ds\widehat{h}(k)$$

$$= (2\pi)^{-1/2}\int_{\mathbb{R}} \mathcal{F}^{-1}M(s)\sum_{k\in\mathbb{Z}}\widehat{h}(k)e^{ik(t-s)}\,ds$$

$$= (2\pi)^{-1/2}\int_{\mathbb{R}} \mathcal{F}^{-1}M(s)h(t-s)\,ds$$

$$= (2\pi)^{-1/2}\mathcal{F}^{-1}M * h(t),$$

where $\mathcal{F}^{-1}M * h$ stands for the convolution of $\mathcal{F}^{-1}M$ and h. The statement follows now directly from Young's inequality. $\square$

The inequality (4.31) is very useful when we try to estimate the norm of functions belonging to the Besov space $B^s_{\infty,\infty}(\mathbb{S}), s > 0$. It will reduce the proof of Theorem (4.5.1) to seeking uniform estimates for the L^1-norm of countably many continuous functions with compact support.

Proof *(Proof of Theorem 4.5.1)* Let $r, s \in \mathbb{R}_{>0}$ and $(M_k) \subset \mathbb{C}$ be a sequence for which the linearisation conditions $(i) - (iii)$ hold. From the definition of the Besov norm it suffices to find a constant $C > 0$ such that

$$2^{rj}\left\|\sum_{k\in\mathbb{Z}} \phi_j(k)M_k\widehat{f}(k)e^{ikt}\right\|_{C(\mathbb{S},\mathbb{C})} \leq C2^{sj}\left\|\sum_{k\in\mathbb{Z}} \phi_j(k)\widehat{f}(k)e^{ikt}\right\|_{C(\mathbb{S},\mathbb{C})}$$

for all $j \in \mathbb{N}$. Consequently, we look for $C > 0$ such that

$$\left\| \sum_{k \in \mathbb{Z}} \phi_j(k) \left(2^{(r-s)j} M_k \right) \widehat{f}(k) e^{ikt} \right\|_{C(\mathbb{S},\mathbb{C})} \leq C \left\| \sum_{k \in \mathbb{Z}} \phi_j(k) \widehat{f}(k) e^{ikt} \right\|_{C(\mathbb{S},\mathbb{C})}$$

for all $j \in \mathbb{N}$ and $f \in \mathcal{D}'(\mathbb{S}, \mathbb{C})$. Given $j \in \mathbb{N}, j > 0$, we define the piecewise affine function $\widetilde{M}_j : \mathbb{R} \to \mathbb{C}$ by

$$\widetilde{M}_j = 0 \quad \text{on} \quad [|t| \leq 2^{j-2}] \cup [|t| \geq 2^{j+2}],$$
$$\widetilde{M}_j(k) = 2^{(r-s)j} M_k \quad \text{for} \quad 2^{j-1} \leq |k| \leq 2^{j+1},$$

and additionally $\widetilde{M}_j$ is affine on each interval $[k, k+1], k \in \mathbb{Z}$. For $j = 0$, the mapping $\widetilde{M}_0$ is defined to be the linear affine function on any interval $[k, k+1] \subset \mathbb{R}$, $k \in \mathbb{Z}$, which satisfies

$$\widetilde{M}_0(k) = \begin{cases} M_k, & \text{for } |k| \leq 2, \\ 0, & \text{for } |k| \geq 4. \end{cases}$$

Taking into consideration that $\operatorname{supp} \phi_j \subset [2^{j-1} \leq |t| \leq 2^{j+1}]$ for all $j \geq 1$ and that $\widetilde{M}_j$ are continuous functions with compact support, we obtain from Lemma 4.5.2 that

$$\left\| \sum_{k \in \mathbb{Z}} \phi_j(k) \left(2^{(r-s)j} M_k \right) \widehat{f}(k) e^{ikt} \right\|_{C(\mathbb{S},\mathbb{C})} = \left\| \sum_{k \in \mathbb{Z}} \phi_j(k) \widetilde{M}_j(k) \widehat{f}(k) e^{ikt} \right\|_{C(\mathbb{S},\mathbb{C})}$$

$$\leq (2\pi)^{-1/2} \| \mathcal{F}^{-1} \widetilde{M}_j \|_{L^1} \left\| \sum_{k \in \mathbb{Z}} \phi_j(k) \widehat{f}(k) e^{ikt} \right\|_{C(\mathbb{S},\mathbb{C})}.$$

A similar inequality is deduced also when $j = 0$. Hence, it suffices to prove that the inverse Fourier transforms $\mathcal{F}^{-1} \widetilde{M}_j$ are uniformly bounded in $L^1(\mathbb{R}, \mathbb{C})$, i.e.

$$\sup_{j \in \mathbb{N}} \| \mathcal{F}^{-1} \widetilde{M}_j \|_{L^1} < \infty. \tag{4.32}$$

Since $s_1 < \infty$, we obtain for $j \in \mathbb{N}$ with $j \geq 1$ that

$$\|\widetilde{M_j}\|_{C(\mathbb{R})} = \max_{2^{j-1} \leq k \leq 2^{j+1}} 2^{(r-s)j}|M_k| \leq s_1 \max_{2^{j-1} \leq k \leq 2^{j+1}} \frac{2^{(r-s)j}}{k^{r-s}}$$

$$\leq s_1 \begin{cases} \dfrac{2^{(r-s)j}}{2^{(j-1)(r-s)}} & \text{if } r-s \geq 0; \\ \dfrac{2^{(r-s)j}}{2^{(j+1)(r-s)}} & \text{if } r-s < 0; \end{cases} = s_1 2^{|r-s|},$$

so that $(\widetilde{M_j})_{j \geq 1}$ is uniformly bounded in $C(\mathbb{R}, \mathbb{C})$, i.e.

$$\sup_{j \geq 1} \|\widetilde{M_j}\|_{C(\mathbb{R})} \leq s_1 2^{|r-s|}. \tag{4.33}$$

Let $N_j : \mathbb{R} \to \mathbb{C}$, $j \in \mathbb{N}$, be the continuous function defined by

$$N_j(t) = \widetilde{M_j}(2^j t) \quad \text{for} \quad t \in \mathbb{R}.$$

We observe that $\operatorname{supp} N_j \subset [4^{-1} \leq |t| \leq 4]$ and that the relation (4.33) is valid for the sequence $(N_j)_{j \geq 1}$ too. Moreover,

$$(2\pi)^{1/2}\|\mathcal{F}^{-1}N_j\|_{C(\mathbb{R})} \leq \int_{|x| \leq 4} |N_j(t)|\,dt \leq 8 \cdot 2^{|r-s|} s_1 \tag{4.34}$$

for all $j \in \mathbb{N}_{>0}$. Well-known properties of the Fourier transform yield

$$\mathcal{F}^{-1}N_j(t) = 2^{-j}(\mathcal{F}^{-1}\widetilde{M_j})(2^{-j}t)$$

for all $j \in \mathbb{N}$, $j > 0$ and $t \in \mathbb{R}$, so that the variable substitution $t = 2^j y$ leads to

$$\|\mathcal{F}^{-1}N_j\|_{L^1} = \int_{\mathbb{R}} |\mathcal{F}^{-1}N_j|(t)\,dt = \int_{\mathbb{R}} 2^j |\mathcal{F}^{-1}N_j|(2^j \theta)\,d\theta = \int_{\mathbb{R}} |\mathcal{F}^{-1}\widetilde{M_j}|(t)\,dt$$

$$= \|\mathcal{F}^{-1}\widetilde{M_j}\|_{L^1}.$$

Thus, instead of showing (4.32) we prove that the sequence $(N_j)_{j \geq 1}$ is uniformly bounded in $L^1(\mathbb{R}, C)$ and that $\widetilde{M_0} \in L^1(\mathbb{R}, \mathbb{C})$.

Integrating by parts twice, we obtain for $t \in \mathbb{R} \setminus \{0\}$ that

$$(2\pi)^{1/2}(\mathcal{F}^{-1}N_j)(t) = 2^{-j} \int_{2^{j-2} \leq |t| \leq 2^{j+2}} \widetilde{M_j}(\theta) e^{i2^{-j}\theta t} \, d\theta$$

$$= 4 \cdot 2^{(r-s)j} M_{2^{j-1}} \frac{e^{it/2} - e^{it/4}}{t^2} - \frac{1}{2} \cdot 2^{(r-s)j} M_{2^{j+1}} \frac{e^{i4t} - e^{i2t}}{t^2}$$

$$+ \frac{1}{2} \cdot 2^{(r-s)j} M_{-2^{j+1}} \frac{e^{-i2t} - e^{-i4t}}{t^2} - 4 \cdot 2^{(r-s)j} M_{-2^{j-1}} \frac{e^{-it/4} - e^{it/2}}{t^2}$$

$$+ \sum_{k=2^{j-1}}^{2^{j+1}-1} 2^j (2^{(r-s)j} M_{k+1} - 2^{(r-s)j} M_k) \frac{e^{i2^{-j}(k+1)t} - e^{i2^{-j}kt}}{t^2}$$

$$+ \sum_{k=-2^{j+1}}^{-2^{j-1}-1} 2^j (2^{(r-s)j} M_{k+1} - 2^{(r-s)j} M_k) \frac{e^{i2^{-j}(k+1)t} - e^{i2^{-j}kt}}{t^2}.$$

For $t \geq 1$, the sum of the first four terms from the right half side of the preceding equality can be bounded by $18 \cdot 2^{|r-s|} s_1 |t|^{-2}$ uniformly in $j \geq 1$, since

$$2^{(r-s)j} |M_{\pm 2^{j \pm 1}}| = \frac{2^{(r-s)j}}{2^{(r-s)(j\pm 1)}} |\pm 2^{(j\pm 1)}|^{r-s} |M_{\pm 2^{j\pm 1}}| \leq 2^{|r-s|} s_1$$

holds for all $j \in \mathbb{N}$ with $j \geq 1$. Rearranging, we have

$$\sum_{k=2^{j-1}}^{2^{j+1}-1} 2^j (2^{(r-s)j} M_{k+1} - 2^{(r-s)j} M_k) \left(e^{i2^{-j}(k+1)t} - e^{i2^{-j}kt} \right) =$$

$$= -\sum_{p=2^{j-1}}^{2^{j+2}-2} 2^j 2^{(r-s)j} (M_{p+2} - 2M_{p+1} + M_p) e^{i2^{-j}(p+1)t}$$

$$+ 2^j 2^{(r-s)j} (M_{2^{j+1}} - M_{2^{j+1}-1}) e^{i2t} - 2^j 2^{(r-s)j} (M_{2^{j-1}+1} - M_{2^{j-1}}) e^{it/2}.$$

Arguing as above, we find that

$$2^j 2^{(r-s)j} |M_{2^{j-1}+1} - M_{2^{j-1}}| \leq \frac{2^{(r-s+1)j}}{(2^{j-1})^{r-s+1}} |2^{j-1}|^{r-s+1} |M_{2^{j-1}+1} - M_{2^{j-1}}|$$

$$\leq 2^{|r-s+1|} s_2,$$

and analogously
$$2^j 2^{(r-s)j}|M_{2^{j+1}} - M_{2^{j+1}-1}| \leq \frac{2^{(r-s+1)j}}{(2^{j+1}-1)^{r-s+1}} s_2 \leq 2^{|r-s+1|} s_2.$$

Moreover,
$$\sum_{p=2^{j-1}}^{2^{j+2}-2} 2^j 2^{(r-s)j}|M_{p+2} - 2M_{p+1} + M_p| = 2^{(r-s+1)j} \sum_{p=2^{j-1}}^{2^{j+2}-2} |M_{p+2} - 2M_{p+1} + M_p|$$

$$= 2^{(r-s+1)j} \sum_{p=2^{j-1}}^{2^{j+2}-2} \frac{1}{p^{r-s+2}} p^{r-s+2}|M_{p+2} - 2M_{p+1} + M_p|$$

$$\leq 2^{(r-s+1)j} s_3 \sum_{p=2^{j-1}}^{2^{j+2}-2} \frac{1}{p^{r-s+2}} \leq 2^{|r-s+2|} s_3.$$

The second sum of the expression for $(2\pi)^{-1/2}(\mathcal{F}^{-1} N_j)(t)$, when $t \neq 0$, can be estimated similarly. Summarising, we found that for $|t| \geq 1$
$$|(2\pi)^{1/2} \mathcal{F}^{-1}(N_j)(t)| \leq \left(18 \cdot 2^{|r-s|} s_1 + 4 \cdot 2^{|r-s+1|} s_2 + 2 \cdot 2^{|r-s+2|} s_3\right) |t|^{-2}$$
and in view of (4.34) we conclude that relation (4.32) is valid iff $\mathcal{F}^{-1}\widetilde{M}_0$ belongs to $L_1(\mathbb{R}, \mathbb{C})$. The mapping $\widetilde{M}_0$ is defined by

$$\widetilde{M}_0(t) = \begin{cases} M_{-2}(4+t)/2, & -4 \leq t \leq -2, \\ M_{k+1}(t-k) + M_k(k+1-t), & k \leq t \leq k+1, -2 \leq k \leq 1, \\ M_2(4-t)/2, & 2 \leq t \leq 4, \end{cases}$$

so that for $x \neq 0$ we obtain
$$(2\pi)^{-1/2} \mathcal{F}^{-1} \widetilde{M}_0(t) = \frac{M_{-2}}{2} \frac{e^{-i2t} - e^{-i4t}}{t^2} + \sum_{k=-2}^{1} (M_{k+1} - M_k) \frac{e^{i(k+1)t} - e^{ikt}}{t^2}$$
$$- \frac{M_2}{2} \frac{e^{i4t} - e^{i2t}}{t^2}.$$

Taking into consideration the following estimates
$$(2\pi)^{1/2}|\mathcal{F}^{-1}\widetilde{M}_0(t)| \leq \left(18 \max_{-2 \leq k \leq 2} |M_k|\right) t^{-2} \quad \text{for} \quad t \neq 0,$$
$$(2\pi)^{1/2}|\mathcal{F}^{-1}\widetilde{M}_0(t)| \leq 8 \max_{-2 \leq k \leq 2} |M_k| \quad \text{for all} \quad t \in \mathbb{R},$$

we obtain the desired result. □

4.6 Local well-posedness

In this section we prove the existence and uniqueness result stated in Theorem 4.2.1. We begin by showing that $-A_1 \in \mathcal{H}(h^{4+\alpha}(\mathbb{S}), h^{1+\alpha}(\mathbb{S}))$, that is the unbounded operator A_1, defined in Theorem 4.4.1, with definition domain $h^{4+\alpha}(\mathbb{S})$ generates a strongly continuous and analytic semigroup in $\mathcal{L}(h^{1+\alpha}(\mathbb{S}))$, for every $A, G \in \mathbb{R}$. We state:

Theorem 4.6.1 *Given $\alpha > 0$, we have that*
$$-A_1 \in \mathcal{H}(h^{4+\alpha}(\mathbb{S}), h^{1+\alpha}(\mathbb{S})).$$

Having established this result, the proof of Theorem 4.2.1 follows then from the local existence, uniqueness, and regularity results for parabolic problems presented in [45]. The proof of Theorem 4.6.1 is rather involved and needs some preliminary results. In the following we consider the complex Banach spaces $h^{m+\alpha}(\mathbb{S}, \mathbb{C}) = h^{m+\alpha}(\mathbb{S}) + i h^{m+\alpha}(\mathbb{S})$, $m = 1, 4$, and show that the complexification of A_1, which we denote again by A_1, satisfies $-A_1 \in \mathcal{H}(h^{4+\alpha}(\mathbb{S}, \mathbb{C}), h^{1+\alpha}(\mathbb{S}, \mathbb{C}))$. The restriction to $h^{1+\alpha}(\mathbb{S})$ of the analytic semigroup generated by the complexification of A_1 in $\mathcal{L}(h^{1+\alpha}(\mathbb{S}, \mathbb{C}))$, is then an analytic semigroup in $\mathcal{L}(h^{1+\alpha}(\mathbb{S}))$, with A_1 as generator.

Using the same notations as in [5], we have $h^{4+\alpha}(\mathbb{S}, \mathbb{C}) \stackrel{d}{\hookrightarrow} h^{1+\alpha}(\mathbb{S}, \mathbb{C})$ and, given $\kappa_1 \geq 1$ and $\omega > 0$, we write
$$-A_1 \in \mathcal{H}(h^{4+\alpha}(\mathbb{S}, \mathbb{C}), h^{1+\beta}(\mathbb{S}, \mathbb{C}), \kappa_1, \omega)$$
if $\omega - A_1 \in \mathcal{L}is(h^{4+\alpha}(\mathbb{S}, \mathbb{C}), h^{1+\alpha}(\mathbb{S}, \mathbb{C}))$ and
$$\kappa_1^{-1} \leq \frac{\|(\lambda - A_1)\rho\|_{C^{1+\alpha}(\mathbb{S}, \mathbb{C})}}{|\lambda| \|\rho\|_{C^{1+\alpha}(\mathbb{S}, \mathbb{C})} + \|\rho\|_{C^{4+\alpha}(\mathbb{S}, \mathbb{C})}} \leq \kappa_1, \quad \rho \in h^{4+\alpha}(\mathbb{S}, \mathbb{C}) \setminus \{0\}, \text{ Re } \lambda \geq \omega.$$

By [5, Theorem 1.2.2]
$$\mathcal{H}(h^{4+\alpha}(\mathbb{S}, \mathbb{C}), h^{1+\alpha}(\mathbb{S}, \mathbb{C})) = \bigcup_{\substack{\kappa_1 \geq 1 \\ \omega > 0}} \mathcal{H}(h^{4+\alpha}(\mathbb{S}, \mathbb{C}), h^{1+\alpha}(\mathbb{S}, \mathbb{C}), \kappa_1, \omega),$$

so that it suffices to prove that $-A_1 \in \mathcal{H}(h^{4+\alpha}(\mathbb{S},\mathbb{C}), h^{1+\alpha}(\mathbb{S},\mathbb{C})), \kappa_1, \omega)$, for some $\kappa_1 \geq 1$ and $\omega > 0$. In view of [5, Remark 1.2.1 (a)], we are done if we find $\kappa_1 \geq 1$ and $\omega > 0$ such that

$$\lambda - A_1 \in \mathcal{L}is(h^{4+\alpha}(\mathbb{S},\mathbb{C}), h^{1+\alpha}(\mathbb{S},\mathbb{C})), \tag{4.35}$$

$$|\lambda| \cdot \|R(\lambda, A_1)\|_{\mathcal{L}(h^{1+\alpha}(\mathbb{S},\mathbb{C}))} \leq \kappa_1, \tag{4.36}$$

for all $\operatorname{Re} \lambda \geq \omega$. Given λ in the resolvent set $\rho(A_1)$ of A_1, we denote by $R(\lambda, A_1)$ the resolvent operator of A_1.

In order to prove Theorem 4.6.1, we consider more closely the operator A_1 defined in Theorem 4.4.1, and show that it is a Fourier multiplier operator. Therefore, we determine first a Fourier expansion for the function $A_1\rho$. Given $\rho \in h^{4+\alpha}(\mathbb{S})$, recall that

$$A_1\rho = \frac{1}{R^3}\partial_\nu(\Delta,\operatorname{tr})^{-1}(0, \rho''). \tag{4.37}$$

To this scope we consider the Fourier expansion of $\rho = \sum_{k \in \mathbb{Z}} \widehat{\rho}(k)x^k$ and we obtain from the well-known Poisson integral formula that

$$(\Delta, \operatorname{tr})^{-1}(0, \rho'') = \sum_{k \in \mathbb{Z}} -k^2 r^{|k|} \widehat{\rho}(k) x^k$$

for all $r \leq 1$ and $x \in \mathbb{S}$. Taking the derivative with respect to r, in $r = 1$, we conclude

$$A_1\left[\sum_{k \in \mathbb{Z}} \widehat{\rho}(k)x^k\right] = \sum_{k \in \mathbb{Z}} \frac{-|k|^3}{R^3}\widehat{\rho}(k)x^k \tag{4.38}$$

for all $\sum_{k \in \mathbb{Z}} \widehat{\rho}(k)x^k \in h^{4+\alpha}(\mathbb{S})$. For simplicity, let $\lambda_k := -|k|^3/R^3$, $k \in \mathbb{Z}$. Hence, it holds then

$$A_1\left[\sum_{k \in \mathbb{Z}} \widehat{\rho}(k)x^k\right] = \sum_{k \in \mathbb{Z}} \lambda_k \widehat{\rho}(k)x^k \tag{4.39}$$

for all $\rho = \sum_{k \in \mathbb{Z}} \widehat{\rho}(k)x^k \in h^{4+\alpha}(\mathbb{S},\mathbb{C})$. The coefficients λ_k, $k \in \mathbb{Z}$, will play an important role in our further analysis.

For simplicity, we fix

$$\omega = 1. \tag{4.40}$$

The next lemma is very useful further on.

Lemma 4.6.2 *Given $k \in \mathbb{Z}$ and $\lambda \in \mathbb{C}$ with $\operatorname{Re} \lambda \geq \omega$ it holds*

$$|\lambda - \lambda_k| \geq \max\{1, |\lambda|, |\lambda_k|\}.$$

Proof The proof is elementary and we omit it.

$\square$

Let us now consider the operator A_1, given by (4.39), as an operator between Sobolev spaces. Given $\lambda \in \mathbb{C}$ with $\operatorname{Re} \lambda \geq \omega$, the operator $\lambda - A_1$ is an isomorphisms. More exactly, it belongs to $\mathcal{L}is(H^{r+3}(\mathbb{S}, \mathbb{C}), H^r(\mathbb{S}, \mathbb{C}))$ for all $r \geq 0$. This can be seen using the Fourier expansions of the functions belonging to the spaces $H^s(\mathbb{S}, \mathbb{C})$, $s \geq 0$, together with the relation $\lim_{|k| \to \infty}(\lambda_{|k|}/|k|^3) = -1/R^3$ and Lemma 4.6.2. Consequently, for any $\operatorname{Re} \lambda \geq \omega$ and $r \geq 0$, the resolvent $R(\lambda, A_1)$ is a well-defined element of $\mathcal{L}(H^r(\mathbb{S}, \mathbb{C}), H^{r+3}(\mathbb{S}, \mathbb{C}))$. Applying this result we obtain

Proposition 4.6.3 *Let $k \in \{1, 4\}$ and suppose*

$$R(\lambda, A_1) \in \mathcal{L}(C^{1+\alpha}(\mathbb{S}, \mathbb{C}), C^{k+\alpha}(\mathbb{S}, \mathbb{C})),$$

for some $\operatorname{Re} \lambda \geq 1$. Then $R(\lambda, A_1) \in \mathcal{L}(h^{1+\alpha}(\mathbb{S}, \mathbb{C}), h^{k+\alpha}(\mathbb{S}, \mathbb{C}))$.

Proof From the assumption we deduce that

$$R(\lambda, A_1) \in \mathcal{L}(h^{1+\alpha}(\mathbb{S}, \mathbb{C}), C^{4+\alpha}(\mathbb{S}, \mathbb{C})).$$

Given $\rho \in h^{1+\alpha}(\mathbb{S}, \mathbb{C})$, we find a sequence $(\rho_n)_n \subset H^r(\mathbb{S}, \mathbb{C}), r > 3$, such that $\rho_n \to \rho$ in $C^{1+\alpha}(\mathbb{S}, \mathbb{C})$. It follows that

$$R(\lambda, A_1)\rho_n \longrightarrow R(\lambda, A_1)\rho \quad \text{in} \quad C^{k+\alpha}(\mathbb{S}, \mathbb{C}).$$

Thanks to the above mentioned result we obtain

$$R(\lambda, A_1)\rho \in \overline{H^{r+3}(\mathbb{S}, \mathbb{C})}^{\|\cdot\|_{C^{k+\alpha}(\mathbb{S}, \mathbb{C})}} = h^{k+\alpha}(\mathbb{S}, \mathbb{C}).$$

$\square$

This proposition is very useful because it allows us to transfer the problem of studying the spectrum of the operator A_1 considered as an operator between small

Hölder spaces to the case when it acts between Hölder spaces, which is more at hand due to the identification $C^s(\mathbb{S}, \mathbb{C}) = B^s_{\infty,\infty}(\mathbb{S})$ for $s > 0, s \notin \mathbb{N}$.

Proof *(Proof of Theorem 4.6.1)* Let $\operatorname{Re} \lambda \geq \omega$ be fixed. Given $\sum_{k \in \mathbb{Z}} \widehat{\rho}(k) x^k \in L^2(\mathbb{S}, \mathbb{C})$, we have in $H^3(\mathbb{S}, \mathbb{C})$

$$R(\lambda, A_1) \left[\sum_{k \in \mathbb{Z}} \widehat{\rho}(k) x^k \right] = \sum_{k \in \mathbb{Z}} M_k^\lambda \widehat{\rho}(k) x^k,$$

where $M_k^\lambda = 1/(\lambda - \lambda_k)$, $k \in \mathbb{Z}$. We prove first that

$$R(\lambda, A_1) \in \mathcal{L}(C^{1+\alpha}(\mathbb{S}, \mathbb{C}), C^{4+\alpha}(\mathbb{S}, \mathbb{C})).$$

Thanks to Theorem 4.5.1 and Proposition 4.6.3 it suffices to show that the coefficients (M_k^λ) satisfy conditions (i), (ii) and (iii) of the above mentioned theorem, with $s = 1 + \alpha$ and $r = 4 + \alpha$.

The relation $\lim_{|k| \to \infty} k^3/(\lambda - \lambda_k) = R^3$ implies (i). Further on we write for $k \geq 1$

$$k^4 |M_{k+1}^\lambda - M_k^\lambda| = \frac{|k|^3}{|\lambda - \lambda_{k+1}|} \frac{|k|^3}{|\lambda - \lambda_k|} \frac{|\lambda_{k+1} - \lambda_k|}{k^2} \xrightarrow[k \to \infty]{} \frac{3}{R^3},$$

because of $(\lambda_k - \lambda_{k+1})/k^2 \to 3R^3$, and (ii) is proved. We also have

$$|k|^5 |M_{k+2} - 2M_{k+1} + M_k| =$$

$$= \frac{|k|^3}{|\lambda - \lambda_{k+2}|} \frac{|k|^3}{|\lambda - \lambda_{k+1}|} \frac{|k|^3}{|\lambda - \lambda_k|} \frac{1}{k^4} |\lambda(\lambda_{k+2} - 2\lambda_{k+1} + \lambda_k) +$$

$$+ \lambda_k(\lambda_{k+1} - \lambda_{k+2}) + \lambda_{k+2}(\lambda_{k+1} - \lambda_k)|,$$

and $(\lambda_{k+2} - 2\lambda_{k+1} + \lambda_k)/k^4 \to 0$, respectively $(\lambda_k(\lambda_{k+1} - \lambda_{k+2}) + \lambda_{k+2}(\lambda_{k+1} - \lambda_k))/k^4 \to 12R^6$. This proves (iii).

Thus, $\lambda - A_1 \in \mathcal{L}is(C^{4+\alpha}(\mathbb{S}, \mathbb{C}), C^{1+\alpha}(\mathbb{S}, \mathbb{C})$ and due to Proposition 4.6.3 we get $\lambda - A_1 \in \mathcal{L}is(h^{4+\alpha}(\mathbb{S}, \mathbb{C}), h^{1+\alpha}(\mathbb{S}, \mathbb{C}))$ for $\operatorname{Re} \lambda \geq \omega$.

We prove now the remaining estimate (4.36). Denoting by $N_k^\lambda := \lambda M_k^\lambda$ for $\operatorname{Re} \lambda \geq \omega$ and $k \in \mathbb{Z}$, it suffices to show that the conditions (i), (ii) and (iii) of Theorem 4.5.1 (with $s = r = 1 + \alpha$) hold for $(N_k^\lambda)_{k \in \mathbb{Z}}$ uniformly in $\lambda \in \{\lambda :$

$\operatorname{Re}\lambda \geq \omega\}$. This is due to the fact that $\lambda R(\lambda, A_1)$ is also a Fourier multiplier, that is

$$\lambda R(\lambda, A_1) \left[\sum_{k \in \mathbb{Z}} \widehat{\rho}(k) x^k \right] = \sum_{k \in \mathbb{Z}} N_k^\lambda \widehat{\rho}(k) x^k \quad \text{for} \quad \sum_{k \in \mathbb{Z}} \widehat{\rho}(k) x^k \in h^{1+\alpha}(\mathbb{S}).$$

From Lemma 4.6.2 we know that $|\lambda - \lambda_k| \geq \max\{1, |\lambda_k|, |\lambda|\}$ for all $\operatorname{Re}\lambda \geq \omega$ and $k \in \mathbb{Z}$. Therefore we get

$$\sup_{\operatorname{Re}\lambda \geq \omega} \sup_{k \in \mathbb{Z}} |N_k^\lambda| \leq 1.$$

We also have

$$|k||N_{k+1}^\lambda - N_k^\lambda| = \frac{|\lambda|}{|\lambda - \lambda_{k+1}|} \frac{|k|^3}{|\lambda - \lambda_k|} \frac{|\lambda_{k+1} - \lambda_k|}{k^2} \leq R^3 \frac{|\lambda_{k+1} - \lambda_k|}{k^2},$$

and we are left to verify that (iii) holds. For $k \neq 0$ we have

$$k^2 |N_{k+2}^\lambda - 2N_{k+1}^\lambda + N_k^\lambda| =$$

$$= \frac{|\lambda|}{|\lambda - \lambda_{k+2}|} \frac{k^3}{|\lambda - \lambda_{k+1}|} \frac{k^3}{|\lambda - \lambda_k|} \frac{1}{k^4} |\lambda(\lambda_{k+2} - 2\lambda_{k+1} + \lambda_k) +$$

$$+ \lambda_k(\lambda_{k+1} - \lambda_{k+2}) + \lambda_{k+2}(\lambda_{k+1} - \lambda_k)| \leq$$

$$\leq R^2 \frac{|\lambda_{k+2} - 2\lambda_{k+1} + \lambda_k|}{|k|} +$$

$$+ R^4 \frac{|\lambda_k(\lambda_{k+1} - \lambda_{k+2}) + \lambda_{k+2}(\lambda_{k+1} - \lambda_k)|}{k^4}.$$

The relation $(\lambda_{k+2} - 2\lambda_{k+1} + \lambda_k)/k \xrightarrow[k \to \infty]{} -6R^3$ and the previous computations lead to $-A_1 \in H(h^{4+\alpha}(\mathbb{S}, \mathbb{C}), h^{1+\alpha}(\mathbb{S}, \mathbb{C}))$. The desired assertion follows now in view of [45, Corollarly 2.1.3]. $\square$

We come now to the proof of the main result of this chapter. The proof justifies why we chosen to work with the small Hölder spaces. The well-known interpolation property of the small Hölder spaces

$$(h^{\sigma_0}(\mathbb{S}), h^{\sigma_1}(\mathbb{S}))_\theta = h^{(1-\theta)\sigma_0 + \theta\sigma_1}, (\mathbb{S}) \tag{4.41}$$

if $\theta \in (0,1)$ and $(1-\theta)\sigma_0 + \theta\sigma_1 \notin \mathbb{N}$, will ensure that the derivative $\partial\Phi(0)$ generates a strongly continuous and analytic semigroup. Furthermore, these spaces fit very well in the abstract theory for parabolic equations presented in [45], which makes it possible for us to use the general results presented there.

Proof *(Proof of the Theorem 4.2.1)* Since the constant $\alpha \in (0,1)$, fixed at the beginning of this chapter, was arbitrary, we get that the assertions of Theorem 4.4.1 and Theorem 4.6.1 hold with α replaced by β, for some fixed $\beta \in (0,\alpha)$. Particularly, we have that $-A_1 \in \mathcal{H}(h^{4+\beta}(\mathbb{S},\mathbb{C}), h^{1+\beta}(\mathbb{S},\mathbb{C}))$, for some $\beta \in (0,\alpha)$. The definition of A_2 is an interpolation space between $h^{1+\beta}(\mathbb{S})$ and $h^{4+\beta}(\mathbb{S})$, since

$$h^{2+\beta}(\mathbb{S}) = (h^{1+\beta}(\mathbb{S}), h^{4+\beta}(\mathbb{S}))_{\frac{1}{3}}.$$

We infer from [45, Proposition 2.4.1] that the sum In view of [5, Theorem 1.3.1], the set $\mathcal{H}(h^{4+\beta}(\mathbb{S}), h^{1+\beta}(\mathbb{S}))$ is open in $\mathcal{L}(h^{4+\beta}(\mathbb{S}), h^{1+\beta}(\mathbb{S}))$, whence there exists an open neighbourhood $\mathcal{O}_\beta$ of 0 in $h^{4+\beta}(\mathbb{S})$ with the property that $-\partial\Phi(\rho) \in \mathcal{H}(h^{4+\beta}(\mathbb{S}), h^{1+\beta}(\mathbb{S}))$ for all $\rho \in \mathcal{O}_\beta$. Let $\mathcal{O} := \mathcal{O}_\beta \cap h^{4+\alpha}(\mathbb{S})$. Letting $C > 0$ be the norm of the compact embedding $h^{4+\alpha}(\mathbb{S}) \hookrightarrow h^{4+\beta}(\mathbb{S})$, we find that $B_{h^{4+\alpha}(\mathbb{S})}(\rho, r/C) \subset \mathcal{O}$ for any ball $B_{h^{4+\beta}(\mathbb{S})}(\rho, r) \subset \mathcal{O}_\beta$. Thus $\mathcal{O}$ is an open neighbourhood of 0 in $\mathcal{V}$.

Given $\rho \in \mathcal{O}$, the operator $\partial\Phi(\rho)$ is the part in $h^{1+\alpha}(\mathbb{S})$ of the sectorial operator $\partial\Phi(\rho) : h^{4+\beta}(\mathbb{S}) \subset h^{1+\beta}(\mathbb{S}) \to h^{1+\beta}(\mathbb{S})$. This follows from (4.41), since for $\theta = (\alpha - \beta)/3$ we have

$$D_S(\theta) = (h^{1+\beta}(\mathbb{S}), h^{4+\beta}(\mathbb{S}))_\theta = h^{1+\alpha}(\mathbb{S}).$$

Thus, we have establish that the assumptions of Theorem 8.4.1 in [45] hold and the proof of Theorem 4.2.1 is now obvious. Consequently, given $\rho_0 \in \mathcal{O}$, there exists a positive time $T > 0$ and a unique classical solution ρ to problem (4.1) on $[0,T]$ satisfying $\rho([0,T]) \subset \mathcal{O}$. Moreover, the solution may be extended on a maximal interval $[0, T(\rho_0))$ and if ρ is uniformly continuous with values in $h^{4+\alpha}(\mathbb{S})$, then either

$$\lim_{t \nearrow T(\rho_0)} \rho(t) \in \partial\mathcal{O} \quad \text{or} \quad T(\rho_0) = +\infty.$$

This completes the proof of Theorem 4.2.1.

□

Chapter 5

Stability properties

We are going to study here the stability properties of the unique radially symmetric solution determined in Chapter 3. Therefore, we shall assume throughout this chapter that $A \in (0, f(1))$ and $G \neq 0$. In order to study stability properties of equilibria we have to determine the spectrum of the complexification of the Fréchet derivative $\partial\Phi(0)$. Being interested in the evolution of tumors which are initially near the unique circular equilibrium $D(0, R_A)$ we take the constant radius R, fixed at the beginning of Chapter 4, to be $R = R_A$. Particularly, functions which belong to $\mathcal{V}$ parametrise curves which are close to the radially symmetric equilibrium. The stability results established in Theorem 5.3.1 and Theorem 5.5.1 are obtained by applying the principle of linearised stability to particular situations.

However, in the case when $G = 0$ and A arbitrary, situation presented Section 5.4, any circle is a stationary solution of (2.1). We choose in there $R = 1$ and show in Theorem 5.4.3 that the circles near the unitary disc build a three dimensional manifold which attracts at an exponential rate the solutions of (2.1) which are initially close to it.

Repeating the arguments presented in the proofs of Theorem 4.2.1 and 4.6.1 we see that the complexification of $\partial\Phi(0)$ generates a strongly continuous and analytic semigroup. Taking into consideration that the embedding $h^{4+\alpha}(\mathbb{S}, \mathbb{C}) \hookrightarrow h^{1+\alpha}(\mathbb{S}, \mathbb{C})$ is compact, we deduce that the complexification of $\partial\Phi(0)$ has a compact resolvent. From [44, Theorem III.8.29], we conclude that its spectrum consists only of eigenvalues of finite multiplicity,

$$\sigma(\partial\Phi(0)) = \sigma_p(\partial\Phi(0)).$$

Therefore, given $\rho \in h^{4+\alpha}(\mathbb{S})$, we look for the Fourier expansion of $\partial\Phi(0)[\rho]$.

Having shown that $\partial\Phi(0)$ is a Fourier multiplier operator, then the point spectrum of $\partial\Phi(0)$ is exactly the symbol of this multiplier.

5.1 Determining the spectrum of the Fréchet derivative of Φ

Let $R = R_A$. In order to obtain a Fourier expansion for the derivative $\partial\Phi(0)[\rho]$, $\rho \in \mathcal{V}$, we are left to determine the Fréchet derivative in 0 of the solution operator $\mathcal{T}$ defined in Theorem 4.3.5. In view of the relation (4.22) we find then the expansion of $\partial\Phi(0)[\rho]$. The main result of this section is Theorem 5.1.6, which states that $\partial\Phi(0)$ is a Fourier multiplier operator and its symbol is also explicitly determined. However, the computations are more involved when computing $\partial\mathcal{T}(0)$, compared to those in Lemma 4.4.4, since $\mathcal{T}(0)$ is not a constant function. We have that:

Lemma 5.1.1 *Given $\rho \in h^{4+\alpha}(\mathbb{S})$, the function $\partial\mathcal{T}(0)[\rho]$ is the unique solution of the linear Dirichlet problem*

$$\Delta w - R_A^2 f'(v_0) w =$$
$$- v_{0,11} \frac{-2x_2^2 \varphi \rho - 2x_1^2 |x| \varphi' \rho + 2x_1 x_2 \varphi \rho'}{|x|^3}$$
$$- 2v_{0,12} \frac{2x_1 x_2 \varphi \rho - 2x_1 x_2 |x| \varphi' \rho + (x_2^2 - x_1^2) \varphi \rho'}{|x|^3}$$
$$+ v_{0,22} \frac{2x_1^2 \varphi \rho + 2x_2^2 |x| \varphi' \rho + 2x_1 x_2 \varphi \rho'}{|x|^3}$$
$$- v_{0,1} \frac{x_1 \varphi \rho - x_1 |x| \varphi' \rho + 2x_2 \varphi \rho' - x_1 |x|^2 \varphi'' \rho - x_1 \varphi \rho''}{|x|^3}$$
$$- v_{0,2} \frac{x_2 \varphi \rho - x_2 |x| \varphi' \rho - 2x_1 \varphi \rho' - x_2 |x|^2 \varphi'' \rho - x_2 \varphi \rho''}{|x|^3} \quad \text{in } \Omega,$$
$$w = 0 \quad \text{on } \mathbb{S}, \tag{5.1}$$

where $v_0 = \mathcal{T}(0)$.

Proof Given $\rho \in \mathcal{V}$, we let $w = w(\rho)$ denote the solution to (5.1). We want to prove the following statement

$$\frac{\|\mathcal{T}(\rho) - \mathcal{T}(0) - w\|_{BUC^{2+\alpha}(\Omega)}}{\|\rho\|_{C^{4+\alpha}(\mathbb{S})}} \xrightarrow[\rho \to 0]{} 0. \tag{5.2}$$

Let $z := \mathcal{T}(\rho) - \mathcal{T}(0) - w \in buc^{2+\alpha}(\Omega)$ and denote by $\mathcal{F}(\rho, v_0)$ the right hand side of the first equation of (5.1). As we mentioned before, $\mathcal{A}(0) = R_A^{-2}\Delta$. Consequently, we get that

$$\mathcal{A}(\rho)\mathcal{T}(\rho) - f(\mathcal{T}(\rho)) - (\mathcal{A}(0)\mathcal{T}(0) - f(\mathcal{T}(0)))$$
$$- \left(\mathcal{A}(0)w - f'(v_0)w - \frac{1}{R_A^2}\mathcal{F}(\rho, v_0)\right) = 0 \quad \text{in} \quad \Omega.$$

It follows that z is the unique solution of the Dirichlet problem

$$\mathcal{A}(\rho)z - f'(v_0)z = \underbrace{f(v) - f(v_0) - f'(v_0)(v - v_0)}_{O\left(\|\rho\|_{C^{4+\alpha}(S)}^2\right)}$$

$$- \underbrace{w_{ij}(b_{ij}(\rho) - b_{ij}(0)) - w_i(b_i(\rho) - b_i(0))}_{O\left(\|\rho\|_{C^{4+\alpha}(S)}^2\right)}$$

$$- v_{0,ij}(b_{ij}(\rho) - b_{ij}(0)) - v_{0,i}(b_i(\rho) - b_i(0))$$
$$- R_A^{-2}\mathcal{F}(\rho, v_0) \quad \text{in} \quad \Omega,$$
$$z = 0 \quad \text{on} \quad \mathbb{S}.$$

We are left to prove that

$$v_{0,ij}(b_{ij}(\rho) - b_{ij}(0)) + v_{0,i}(b_i(\rho) - b_i(0)) + \frac{1}{R_A^2}\mathcal{F}(\rho, v_0) = O\left(\|\rho\|_{C^{4+\alpha}(S)}^2\right) \quad (5.3)$$

for ρ sufficiently small. Using the estimates (3.12) and (6.36) in [37], as we did in the proof of Theorem 4.4.4, we obtain the desired result. For $x \neq 0$ we have the following relations:

The coefficient of $v_{0,11}$ in (5.3)

$$b_{11}(\rho) - b_{11}(0) - \left(\frac{-2x_2^2\varphi\rho - 2x_1^2|x|\varphi'\rho + 2x_1x_2\varphi\rho'}{R_A^2|x|^3}\right) =$$

$$= (f_{\rho,1}(\Theta_\rho) - f_{\rho,1}(\Theta_0))(f_{\rho,1}(\Theta_\rho) + f_{\rho,1}(\Theta_0))$$

$$+ f_{\rho,2}^2(\Theta_\rho) - \left(\frac{-2x_2^2\varphi\rho - 2x_1^2|x|\varphi'\rho + 2x_1x_2\varphi\rho'}{R_A^2|x|^3}\right)$$

$$= O\left(\|\rho\|^2_{C^{4+\alpha}(S)}\right) + \frac{1}{R_A^2(1+\varphi'\rho)^2} - \frac{1}{R^2}$$

$$+ \frac{2}{R_A^2} \frac{x_2(-x_2\varphi\rho + x_2|x|\varphi'\rho + x_1\varphi\rho')}{|x|^3\left(1 + \frac{1}{|x|}\varphi\rho\right)(1+\varphi'\rho)}$$

$$- \left(\frac{-2x_2^2\varphi\rho - 2x_1^2|x|\varphi'\rho + 2x_1x_2\varphi\rho'}{R_A^2|x|^3}\right) = O\left(\|\rho\|^2_{C^{4+\alpha}(S)}\right).$$

The coefficient of $v_{0,12}$ in (5.3)

$$b_{12}(\rho) - b_{12}(0) - \frac{2x_1x_2\varphi\rho - 2x_1x_2|x|\varphi'\rho + (x_2^2 - x_1^2)\varphi\rho'}{R_A^2|x|^3} =$$

$$= f_{\rho,1}(\Theta_\rho)g_{\rho,1}(\Theta_\rho) + f_{\rho,2}(\Theta_\rho)g_{\rho,2}(\Theta_\rho)$$

$$- \frac{2x_1x_2\varphi\rho - 2x_1x_2|x|\varphi'\rho + (x_2^2 - x_1^2)\varphi\rho'}{R_A^2|x|^3}$$

$$= O\left(\|\rho\|^2_{C^{4+\alpha}(S)}\right) +$$

$$+ \frac{1}{R_A(1+\varphi'\rho)} \frac{x_1x_2\varphi\rho - x_1x_2|x|\varphi'\rho + x_2^2\varphi\rho'}{R_A|x|^3\left(1 + \frac{1}{|x|}\varphi\rho\right)(1+\varphi'\rho)}$$

$$+ \frac{1}{R_A(1+\varphi'\rho)} \frac{x_1x_2\varphi\rho - x_1x_2|x|\varphi'\rho - x_1^2\varphi\rho'}{R_A|x|^3\left(1 + \frac{1}{|x|}\varphi\rho\right)(1+\varphi'\rho)} -$$

$$- \frac{2x_1x_2\varphi\rho - 2x_1x_2|x|\varphi'\rho + (x_2^2 - x_1^2)\varphi\rho'}{R_A^2|x|^3} = O\left(\|\rho\|^2_{C^{4+\alpha}(S)}\right).$$

The coefficient of $v_{0,22}$ in (5.3)

$$b_{22}(\rho) - b_{22}(0) + \frac{2x_1^2 \varphi \rho + 2x_2^2 |x| \varphi' \rho + 2x_1 x_2 \varphi \rho'}{R_A^2 |x|^3} =$$

$$= (g_{\rho,2}(\Theta_\rho) - g_{\rho,2}(\Theta_0))(g_{\rho,2}(\Theta_\rho) + g_{\rho,2}(\Theta_0))$$

$$+ g_{\rho,1}^2(\Theta_\rho) + \frac{2x_1^2 \varphi \rho + 2x_2^2 |x| \varphi' \rho + 2x_1 x_2 \varphi \rho'}{R_A^2 |x|^3}$$

$$= O\left(\|\rho\|_{C^{4+\alpha}(\mathbb{S})}^2\right) + \frac{1}{R_A^2(1+\varphi'\rho)^2} - \frac{1}{R^2}$$

$$+ \frac{2}{R_A^2} \frac{-x_1^2 \varphi \rho + x_1^2 |x| \varphi' \rho - x_1 x_2 \varphi \rho'}{|x|^3 \left(1 + \frac{1}{|x|}\varphi\rho\right)(1+\varphi'\rho)}$$

$$+ \frac{2x_1^2 \varphi \rho + 2x_2^2 |x| \varphi' \rho + 2x_1 x_2 \varphi \rho'}{R_A^2 |x|^3} = O\left(\|\rho\|_{C^{4+\alpha}(\mathbb{S})}^2\right).$$

In order to estimate the $BUC^\alpha(\Omega)$ norm of the coefficients of $v_{0,i}$ in (5.3) we have to determine $f_{\rho,11}(\Theta_\rho)$, $g_{\rho,11}(\Theta_\rho)$, $f_{\rho,22}(\Theta_\rho)$, and $g_{\rho,22}(\Theta_\rho)$, respectively. Since $f_{0,11}(\Theta_0) = g_{0,11}(\Theta_0) = f_{0,22}(\Theta_0) = g_{0,22}(\Theta_0) = 0$ we are left to specify only the terms of first order in ρ of these second order derivatives. As we see below, the derivatives contain no zero order terms in ρ. The chain rule yields

$$\begin{bmatrix} f_{\rho,11}(\Theta_\rho) & f_{\rho,12}(\Theta_\rho) \end{bmatrix} \partial \Theta_\rho = \begin{bmatrix} \partial_1(f_{\rho,1}(\Theta_\rho)) & \partial_2(f_{\rho,1}(\Theta_\rho)) \end{bmatrix},$$

thus

$$\begin{bmatrix} f_{\rho,11}(\Theta_\rho) & f_{\rho,12}(\Theta_\rho) \end{bmatrix} = \begin{bmatrix} \partial_1(f_{\rho,1}(\Theta_\rho)) & \partial_2(f_{\rho,1}(\Theta_\rho)) \end{bmatrix} \partial \Psi_\rho(\Theta_\rho).$$

Since

$$\partial \Psi_\rho(\Theta_\rho) = \begin{bmatrix} f_{\rho,1}(\Theta_\rho) & f_{\rho,2}(\Theta_\rho) \\ g_{\rho,1}(\Theta_\rho) & g_{\rho,2}(\Theta_\rho) \end{bmatrix},$$

we obtain that

$$f_{\rho,11}(\Theta_\rho) = \partial_1(f_{\rho,1}(\Theta_\rho))f_{\rho,1}(\Theta_\rho) + \partial_2(f_{\rho,1}(\Theta_\rho))g_{\rho,1}(\Theta_\rho) =$$

$$= \frac{1}{R_A^2(1+\varphi'\rho)} \left\{ \frac{-x_1|x|\varphi''\rho + x_2\varphi'\rho'}{|x|^2(1+\varphi'\rho)^2} + \right.$$

$$+ \frac{1}{|x|^5(1+\varphi'\rho)\left(1+\frac{1}{|x|}\varphi\rho\right)} [3x_1 x_2^2 \varphi\rho - 3x_1 x_2^2 |x|\varphi'\rho$$

$$+ (x_2^3 + x_2|x|^2 - 3x_1^2 x_2)\varphi\rho' + (x_1^2 x_2|x| - x_2^3|x|)\varphi'\rho' +$$

$$\left. + x_1 x_2^2 |x|^2 \varphi''\rho - x_1 x_2^2 \varphi\rho''] \right\} + O\left(\|\rho\|_{C^{4+\alpha}(\mathbb{S})}^2\right).$$

Similarly, we have
$$\begin{bmatrix} f_{\rho,21}(\Theta_\rho) & f_{\rho,22}(\Theta_\rho) \end{bmatrix} \partial\Theta_\rho = \begin{bmatrix} \partial_1(f_{\rho,2}(\Theta_\rho)) & \partial_2(f_{\rho,2}(\Theta_\rho)) \end{bmatrix},$$
thus
$$f_{\rho,22}(\Theta_\rho) = \partial_1(f_{\rho,2}(\Theta_\rho))f_{\rho,2}(\Theta_\rho) + \partial_2(f_{\rho,2}(\Theta_\rho))g_{\rho,2}(\Theta_\rho) =$$

$$= \frac{1}{R_A^2(1+\varphi'\rho)^2\left(1+\frac{1}{|x|}\varphi\rho\right)|x|^5} \left[(x_1|x|^2 - 3x_1 x_2^2)\varphi\rho + \right.$$

$$+ (3x_1 x_2^2 |x| - x_1|x|^3)\varphi'\rho(x_1^2 x_2 + 3x_1^2 x_2)\varphi\rho' - 2x_1^2 x_2|x|\varphi'\rho'$$

$$\left. - x_1 x_2^2 |x|^2 \varphi''\rho - x_1^3 \varphi\rho''\right] + O\left(\|\rho\|_{C^{4+\alpha}(\mathbb{S})}^2\right).$$

The coefficient of $v_{0,1}$ in (5.3)
$$b_1(\rho) - b_1(0) - \frac{x_1\varphi\rho - x_1|x|\varphi'\rho + 2x_2\varphi\rho' - x_1|x|^2\varphi''\rho - x_1\varphi\rho''}{R_A^2|x|^3}$$

$$= f_{\rho,11}(\Theta_\rho) + f_{\rho,22}(\Theta_\rho)$$

$$- \frac{x_1\varphi\rho - x_1|x|\varphi'\rho + 2x_2\varphi\rho' - x_1|x|^2\varphi''\rho - x_1\varphi\rho''}{R_A^2|x|^3}$$

$$= O\left(\|\rho\|^2_{C^{4+\alpha}(S)}\right) + \frac{1}{R_A^2(1+\varphi'\rho)}\left\{\frac{-x_1|x|\varphi''\rho + x_2\varphi'\rho'}{|x|^2(1+\varphi'\rho)^2} + \right.$$

$$+ \frac{1}{|x|^5(1+\varphi'\rho)\left(1+\frac{1}{|x|}\varphi\rho\right)}[3x_1x_2^2\varphi\rho - 3x_1x_2^2|x|\varphi'\rho +$$

$$+ (x_2^3 + x_2|x|^2 - 3x_1^2x_2)\varphi\rho' + (x_1^2x_2|x| - x_2^3|x|)\varphi'\rho'$$

$$\left. + x_1x_2^2|x|^2\varphi''\rho - x_1x_2^2\varphi\rho'']\right\}$$

$$+ \frac{1}{R_A^2(1+\varphi'\rho)^2\left(1+\frac{1}{|x|}\varphi\rho\right)|x|^5}[(x_1|x|^2 - 3x_1x_2^2)\varphi\rho +$$

$$+ (3x_1x_2^2|x| - x_1|x|^3)\varphi'\rho - 2x_1^2x_2|x|\varphi'\rho'$$

$$+ (x_1^2x_2 + 3x_1^2x_2)\varphi\rho' - x_1x_2^2|x|^2\varphi''\rho - x_1^3\varphi\rho'']$$

$$- \frac{x_1\varphi\rho - x_1|x|\varphi'\rho + 2x_2\varphi\rho' - x_1|x|^2\varphi''\rho - x_1\varphi\rho''}{R_A^2|x|^3}$$

$$= O\left(\|\rho\|^2_{C^{4+\alpha}(S)}\right).$$

To show that the coefficient of $v_{0,2}$ is of second order in ρ, we determine now the first order terms of the second order derivatives $g_{\rho,11}$ and $g_{\rho,22}$. Computations similar to those presented above yield that

$$\begin{bmatrix} g_{\rho,11}(\Theta_\rho) & g_{\rho,12}(\Theta_\rho) \end{bmatrix} = \begin{bmatrix} \partial_1(g_{\rho,1}(\Theta_\rho)) & \partial_2(g_{\rho,1}(\Theta_\rho)) \end{bmatrix} \partial\Psi_\rho(\Theta_\rho),$$

hence

$$g_{\rho,11}(\Theta_\rho) = \partial_1(g_{\rho,1}(\Theta_\rho))f_{\rho,1}(\Theta_\rho) + \partial_2(g_{\rho,1}(\Theta_\rho))g_{\rho,1}(\Theta_\rho) =$$

$$= \frac{1}{R_A^2(1+\varphi'\rho)^2\left(1+\frac{1}{|x|}\varphi\rho\right)|x|^5}\left[(x_2|x|^2 - 3x_1^2x_2)\varphi\rho - x_2^3\varphi\rho''\right.$$

$$+ (3x_1^2 x_2 |x| - x_2 |x|^3)\varphi'\rho - (x_1 x_2^2 + 3x_1 x_2^2)\varphi\rho' + 2x_1 x_2^2 |x|\varphi'\rho'$$
$$- x_1^2 x_2 |x|^2 \varphi''\rho] + O\left(\|\rho\|_{C^{4+\alpha}(S)}^2\right),$$

respectively

$$g_{\rho,22}(\Theta_\rho) = \partial_1(g_{\rho,2}(\Theta_\rho))f_{\rho,2}(\Theta_\rho) + \partial_2(g_{\rho,2}(\Theta_\rho))g_{\rho,2}(\Theta_\rho) =$$

$$= \frac{1}{R_A^2(1+\varphi'\rho)}\left\{\frac{-x_2|x|\varphi''\rho - x_1\varphi'\rho'}{|x|^2(1+\varphi'\rho)^2}\right.$$

$$+ \frac{1}{|x|^5(1+\varphi'\rho)\left(1+\frac{1}{|x|}\varphi\rho\right)}[3x_1^2 x_2 \varphi\rho - 3x_1^2 x_2 |x|\varphi'\rho$$

$$- (x_1^3 + x_1|x|^2 - 3x_1 x_2^2)\varphi\rho' - (x_1 x_2^2|x| - x_1^3|x|)\varphi'\rho'$$

$$\left. + x_1^2 x_2 |x|^2 \varphi''\rho - x_1^2 x_2 \varphi\rho'']\right\} + O\left(\|\rho\|_{C^{4+\alpha}(S)}^2\right).$$

The coefficient of $v_{0,2}$ in (5.3)

$$b_2(\rho) - b_2(0) - \frac{x_2\varphi\rho - x_2|x|\varphi'\rho - 2x_1\varphi\rho' - x_2|x|^2\varphi''\rho - x_2\varphi\rho''}{R_A^2|x|^3} =$$

$$= g_{\rho,11}(\Theta_\rho) + g_{\rho,22}(\Theta_\rho)$$

$$- \frac{x_2\varphi\rho - x_2|x|\varphi'\rho - 2x_1\varphi\rho' - x_2|x|^2\varphi''\rho - x_2\varphi\rho''}{R_A^2|x|^3} =$$

$$= \frac{1}{R_A^2(1+\varphi'\rho)^2\left(1+\frac{1}{|x|}\varphi\rho\right)|x|^5}\left[(x_2|x|^2 - 3x_1^2 x_2)\varphi\rho + 2x_1 x_2^2|x|\varphi'\rho'\right.$$

$$+ (3x_1^2 x_2 |x| - x_2|x|^3)\varphi'\rho - (x_1 x_2^2 + 3x_1 x_2^2)\varphi\rho' - x_1^2 x_2 |x|^2 \varphi''\rho$$

$$-x_2^3\varphi\rho''] + \frac{1}{R_A^2(1+\varphi'\rho)}\left\{\frac{-x_2|x|\varphi''\rho - x_1\varphi'\rho'}{|x|^2(1+\varphi'\rho)^2}\right.$$

$$+ \frac{1}{|x|^5(1+\varphi'\rho)\left(1+\frac{1}{|x|}\varphi\rho\right)}[3x_1^2x_2\varphi\rho - 3x_1^2x_2|x|\varphi'\rho + x_1^2x_2|x|^2\varphi''\rho$$

$$\left.-(x_1^3 + x_1|x|^2 - 3x_1x_2^2)\varphi\rho' - (x_1x_2^2|x| - x_1^3|x|)\varphi'\rho' - x_1^2x_2\varphi\rho'']\right\}$$

$$- \frac{x_2\varphi\rho - x_2|x|\varphi'\rho - 2x_1\varphi\rho' - x_2|x|^2\varphi''\rho - x_2\varphi\rho''}{R_A^2|x|^3} + O\left(\|\rho\|_{C^{4+\alpha}(S)}^2\right)$$

$$= O\left(\|\rho\|_{C^{4+\alpha}(S)}^2\right),$$

and the proof is completed.

$\square$

The result of this lemma is not very useful yet. This is due to the fact that the first equation of (5.1) contains besides ρ and w also derivatives of φ and f. Therefore, it is difficult to determine the Fourier expansion of w, the solution of (5.1), when knowing that of ρ. That is why we formally linearise the free boundary problem describing the stationary states of the full system (2.1) at the unique radially symmetric solution (R_A, ψ_A, p_A), found in Theorem 3.0.2, where we simply write $\psi_A := \psi_{R_A}$ and $p_A := p_{R_A}$. By doing this we shall find a nice decomposition of the derivative $\partial T(0)$ as a sum of two operators (see Lemma 5.1.2 below).

Consider now perturbations of the radially symmetric solution of the form

$$\begin{cases} \psi_\varepsilon &= \psi_A + \varepsilon\psi, \\ p_\varepsilon &= p_A + \varepsilon p, \\ \Omega_{\rho_\varepsilon} &= \{re^{is} : 0 \leq r < R_A(1+\varepsilon\rho(s)), s \in \mathbb{R}\}, \end{cases}$$

where we simply write $\rho(s) = \rho(e^{is})$ for all $s \in \mathbb{R}$. Here, ε is a small parameter, and ψ, p, ρ are unknown functions. Letting $\Delta\psi_\varepsilon = f(\psi_\varepsilon)$ in $\Omega_{\rho_\varepsilon}$ it follows that

$$\Delta\psi_A + \varepsilon\Delta\psi = f(\psi_A + \varepsilon\psi).$$

We subtract $\Delta \psi_A = f(\psi_A)$ from this equation, divide it by ε, and let $\varepsilon \to 0$ to come to
$$\Delta \psi = f'(\psi_A) \cdot \psi \quad \text{in } D(0, R_A). \tag{5.4}$$
Analogously, the linearisation of the second equation of (2.1), is the following relation
$$\Delta p = 0, \quad \text{in } D(0, R_A). \tag{5.5}$$
It is also easy to see that the linearisation of the first boundary condition in (2.1) is given by
$$\psi + R_A \psi'_A(R_A) \rho = 0, \quad \text{on } \partial D(0, R_A). \tag{5.6}$$
In order to prove (5.6) we let $\psi_\varepsilon = 1$ on $\partial \Omega_{\rho_\varepsilon}$. In view of this relation, we get that
$$\psi_A(R_A(1 + \varepsilon\rho(s))e^{is}) + \varepsilon\psi(R_A(1 + \varepsilon\rho(s))e^{is}) = \psi_A(R_A).$$
We substract $\psi_A(R_A)$, divide the difference by ε, and for ε going to zero we come to relation (5.6).

To find the linearisation of the second boundary condition of (2.1), we set $p_\varepsilon = \kappa(\rho_\varepsilon) - AGR_A^2|1+\varepsilon\rho|^2/4$ on the boundary $\partial\Omega_{\rho_\varepsilon}$. We find that
$$p(R_A(1+\varepsilon\rho(s))e^{is}) = \frac{\kappa(\varepsilon\rho)(s) - \kappa(0)}{\varepsilon} - \frac{AGR_A^2}{4} \frac{(1+\varepsilon\rho(s))^2 - 1}{\varepsilon},$$
and for $\varepsilon \to 0$, we get
$$p = -\left(\frac{1}{R_A} + \frac{AGR_A^2}{2}\right)\rho - \frac{1}{R_A}\rho'', \quad \text{on } \partial D(0, R_A). \tag{5.7}$$
Lastly, we linearise the coupling equation. Therefore we set
$$G\langle \nabla N_{\varepsilon\rho}(\Theta_{\varepsilon\rho}), \nabla \psi_\varepsilon(\Theta_{\varepsilon\rho})\rangle - \langle \nabla N_{\varepsilon\rho}(\Theta_{\varepsilon\rho}), \nabla p_\varepsilon(\Theta_{\varepsilon\rho})\rangle$$
$$- \frac{AG}{2}\langle \nabla N_{\varepsilon\rho}(\Theta_{\varepsilon\rho}), \Theta_{\varepsilon\rho}\rangle = 0 \quad \text{on } \mathbb{S}.$$
In view of (4.23), we have $\langle \nabla N_{\varepsilon\rho}(\Theta_{\varepsilon\rho}), \Theta_{\varepsilon\rho}\rangle = R_A(1+\varepsilon\rho)$. Since (R_A, ψ_A, p_A) is the stationary solution of (2.1) we also have that
$$G\langle \nabla N_0(\Theta_0), \nabla \psi_A\rangle - \frac{AGR_A}{2} = 0 \quad \text{on } \mathbb{S}.$$

Subtracting this equation from the previous one, we come, after diving the difference by ε to the following relation

$$G\langle \nabla N_{\varepsilon\rho}(\Theta_{\varepsilon\rho}), \nabla\psi(\Theta_{\varepsilon\rho})\rangle + G\left\langle \frac{\nabla N_{\varepsilon\rho}(\Theta_{\varepsilon\rho}) - \nabla N_0(\Theta_0)}{\varepsilon}, \nabla\psi_A(\Theta_{\varepsilon\rho})\right\rangle$$

$$- \langle \nabla N_{\varepsilon\rho}(\Theta_{\varepsilon\rho}), \nabla p(\Theta_{\varepsilon\rho})\rangle - \frac{AGR_A}{2}\frac{(1+\varepsilon\rho)-1}{\varepsilon} = 0 \quad \text{on } \mathbb{S}.$$

Since ψ_A is radially symmetric, we get that

$$\left\langle \frac{\nabla N_{\varepsilon\rho}(\Theta_{\varepsilon\rho}) - \nabla N_0(\Theta_0)}{\varepsilon}, \nabla\psi_A(\Theta_{\varepsilon\rho})\right\rangle \xrightarrow[\varepsilon\to 0]{} \langle(-x_2, x_1), \nabla\psi_A\rangle\rho' = 0.$$

Whence, for $\varepsilon \to 0$ in the second equation above, we arrive at

$$G\partial_\nu\psi - \partial_\nu p - \frac{AGR_A}{2}\rho = 0, \quad \text{on } \partial D(0, R_A). \tag{5.8}$$

Summarising, the linearisation of problem (2.1) is the free boundary problem consisting of the equations (5.4)-(5.8)

$$\begin{cases} \Delta\psi = f'(\psi_A)\cdot\psi & \text{in } D(0, R_A), \\ \Delta p = 0 & \text{in } D(0, R_A), \\ \psi = -R_A\psi'_A(R_A)\rho & \text{on } \partial D(0, R_A), \\ p = -\left(\frac{1}{R_A} + \frac{AGR_A^2}{2}\right)\rho - \frac{1}{R_A}\rho'' & \text{on } \partial D(0, R_A), \\ G\partial_\nu\psi - \partial_\nu p = \frac{AGR_A}{2}\rho & \text{on } \partial D(0, R_A). \end{cases} \tag{5.9}$$

We look now for a connection between the linearisation (5.9) and the Fréchet derivative of Φ in 0. To this scope, we transform first the system (5.9) to the unitary disc using the Hanzawa diffeomorphism Θ_0, i.e. we set

$$w(x) = \psi(R_A x), \ z(x) = p(R_A x), \ v_0(x) = \psi_A(R_A x),$$

for $x \in \Omega$, and substitute these expressions in (5.9).

This leads to the following system of equations

$$\begin{cases} \Delta w = R_A^2 f'(v_0) \cdot w & \text{in } \Omega, \\ w = -\partial_\nu v_0 \rho & \text{on } \mathbb{S}, \\ \Delta z = 0 & \text{in } \Omega, \\ z = -\left(\dfrac{1}{R_A} + \dfrac{AGR_A^2}{2}\right)\rho - \dfrac{1}{R_A}\rho'' & \text{on } \mathbb{S}, \\ \dfrac{G}{R_A}\partial_\nu \psi - \dfrac{1}{R_A}\partial_\nu p = \dfrac{AGR_A}{2}\rho & \text{on } \mathbb{S}. \end{cases} \quad (5.10)$$

Given $\rho \in h^{4+\alpha}(\mathbb{S})$, we let $\mathcal{W}(\rho) \in buc^{2+\alpha}(\Omega)$ denote the solution to the linear Dirichlet problem

$$\begin{cases} \Delta w = R_A^2 f'(v_0) \cdot w & \text{in } \Omega, \\ w = -\partial_\nu v_0 \rho & \text{on } \mathbb{S}. \end{cases} \quad (5.11)$$

Further on, we want to determine a relation between $\partial \mathcal{T}(0)[\rho]$ and $\mathcal{W}(\rho)$. Therefore, we define the extension operator $\mathcal{E} : h^{4+\alpha}(\mathbb{S}) \to buc^{2+\alpha}(\Omega)$ by

$$\mathcal{E}(\rho)(x) := \left(v_{0,1}(x)\dfrac{x_1}{|x|} + v_{0,2}(x)\dfrac{x_2}{|x|}\right) \rho\left(\dfrac{x}{|x|}\right) \varphi(|x| - 1) \quad (5.12)$$

Using these operators we can now write $\partial \mathcal{T}(0)$ as the sum of $\mathcal{W}$ and $\mathcal{E}$. This decomposition is very useful because we got rid, in this way, of all the terms from the right hand side of first equation of (5.1). Recall that our goal is to determine the Fourier expansion of $\partial_\nu(\partial \mathcal{T}(0)[\rho])$ when $\rho \in h^{4+\alpha}(\mathbb{S})$. It turns out, cf. Lemma 5.1.3, that $\partial_\nu(\mathcal{E}(\rho))$ is collinear with ρ for all $\rho \in h^{4+\alpha}(\mathbb{S})$. Moreover, using ODE–techniques we are able to determine an expansion for $\mathcal{W}(\rho)$ for all $\rho \in h^{4+\alpha}(\mathbb{S})$, cf. (5.23). Indeed, we have:

Lemma 5.1.2 *It holds that*

$$\partial \mathcal{T}(0) = \mathcal{W} + \mathcal{E}.$$

Proof Let $\rho \in h^{4+\alpha}(\mathbb{S})$ be given. From relation (5.12) we obtain by partial differentiation the following relations:

$$\frac{\partial^2 \mathcal{E}(\rho)}{\partial x_1^2}(x) = \left(v_{0,111} \frac{x_1}{|x|} + v_{0,112} \frac{x_2}{|x|} + 2v_{0,11} \frac{x_2^2}{|x|^3} - v_{0,12} \frac{x_1 x_2}{|x|^3} \right.$$

$$\left. - v_{0,1} \frac{3x_1 x_2^2}{|x|^5} - v_{0,2} \frac{x_2^3 - 2x_1^2 x_2}{|x|^5} \right) \rho \varphi$$

$$+ \frac{2x_1}{|x|^2} \left(v_{0,11} \frac{x_1}{|x|} + v_{0,12} \frac{x_2}{|x|} + v_{0,1} \frac{x_2^2}{|x|^3} - v_{0,2} \frac{x_1 x_2}{|x|^3} \right) \rho \varphi'$$

$$- \frac{2x_2}{|x|^2} \left(v_{0,11} \frac{x_1}{|x|} + v_{0,12} \frac{x_2}{|x|} + v_{0,1} \frac{x_2^2}{|x|^3} - v_{0,2} \frac{x_1 x_2}{|x|^3} \right) \rho' \varphi$$

$$+ \frac{x_2^2}{|x|^4} \left(v_{0,1} \frac{x_1}{|x|} + v_{0,2} \frac{x_2}{|x|} \right) \rho'' \varphi - 2 \frac{x_1 x_2}{|x|^3} \left(v_{0,1} \frac{x_1}{|x|} + v_{0,2} \frac{x_2}{|x|} \right) \rho' \varphi'$$

$$+ \frac{x_1^2}{|x|^2} \left(v_{0,1} \frac{x_1}{|x|} + v_{0,2} \frac{x_2}{|x|} \right) \rho \varphi'' + \frac{2x_1 x_2}{|x|^4} \left(v_{0,1} \frac{x_1}{|x|} + v_{0,2} \frac{x_2}{|x|} \right) \rho' \varphi$$

$$+ \frac{x_2^2}{|x|^3} \left(v_{0,1} \frac{x_1}{|x|} + v_{0,2} \frac{x_2}{|x|} \right) \rho \varphi'$$

and

$$\frac{\partial^2 \mathcal{E}(\rho)}{\partial x_2^2}(x) = \left(v_{0,122} \frac{x_1}{|x|} + v_{0,222} \frac{x_2}{|x|} - 2v_{0,12} \frac{x_1 x_2}{|x|^3} + 2v_{0,22} \frac{x_1^2}{|x|^3} \right.$$

$$\left. - v_{0,1} \frac{x_1^3 - 2x_1 x_2^2}{|x|^5} - v_{0,2} \frac{3x_1^2 x_2}{|x|^5} \right) \rho \varphi$$

$$+ \frac{2x_2}{|x|^2} \left(v_{0,12} \frac{x_1}{|x|} + v_{0,22} \frac{x_2}{|x|} - v_{0,1} \frac{x_1 x_2}{|x|^3} + v_{0,2} \frac{x_1^2}{|x|^3} \right) \rho \varphi'$$

$$+ \frac{2x_1}{|x|^2} \left(v_{0,12} \frac{x_1}{|x|} + v_{0,22} \frac{x_2}{|x|} + v_{0,1} \frac{x_1 x_2}{|x|^3} + v_{0,2} \frac{x_1^2}{|x|^3} \right) \rho' \varphi$$

$$+ \frac{x_1^2}{|x|^4}\left(v_{0,1}\frac{x_1}{|x|} + v_{0,2}\frac{x_2}{|x|}\right)\rho''\varphi + 2\frac{x_1 x_2}{|x|^3}\left(v_{0,1}\frac{x_1}{|x|} + v_{0,2}\frac{x_2}{|x|}\right)\rho'\varphi'$$

$$+ \frac{x_2^2}{|x|^2}\left(v_{0,1}\frac{x_1}{|x|} + v_{0,2}\frac{x_2}{|x|}\right)\rho\varphi'' - \frac{2x_1 x_2}{|x|^4}\left(v_{0,1}\frac{x_1}{|x|} + v_{0,2}\frac{x_2}{|x|}\right)\rho'\varphi$$

$$+ \frac{x_1^2}{|x|^3}\left(v_{0,1}\frac{x_1}{|x|} + v_{0,2}\frac{x_2}{|x|}\right)\rho\varphi'.$$

Thus, we obtain
$$\Delta\mathcal{E}(\rho) = -v_{0,11}\frac{-2x_2^2\varphi\rho - 2x_1^2|x|\varphi'\rho + 2x_1 x_2\varphi\rho'}{|x|^3}$$

$$- 2v_{0,12}\frac{2x_1 x_2\varphi\rho - 2x_1 x_2|x|\varphi'\rho + (x_2^2 - x_1^2)\varphi\rho'}{|x|^3}$$

$$+ v_{0,22}\frac{2x_1^2\varphi\rho + 2x_2^2|x|\varphi'\rho + 2x_1 x_2\varphi\rho'}{|x|^3}$$

$$- v_{0,1}\frac{x_1\varphi\rho - x_1|x|\varphi'\rho + 2x_2\varphi\rho' - x_1|x|^2\varphi''\rho - x_1\varphi\rho''}{|x|^3}$$

$$- v_{0,2}\frac{x_2\varphi\rho - x_2|x|\varphi'\rho - 2x_1\varphi\rho' - x_2|x|^2\varphi''\rho - x_2\varphi\rho''}{|x|^3}$$

$$+ \varphi\rho\left(v_{0,111}\frac{x_1}{|x|} + v_{0,112}\frac{x_2}{|x|} + v_{0,122}\frac{x_1}{|x|} + v_{0,222}\frac{x_2}{|x|}\right).$$

Since $\Delta v_0 = R_A^2 f(v_0)$ in Ω, it follows
$$v_{0,111} + v_{0,122} = R_A^2 f'(v_0)\, v_{0,1}, \qquad (5.13)$$

$$v_{0,112} + v_{0,222} = R_A^2 f'(v_0)\, v_{0,2}. \qquad (5.14)$$

Multiplying relation (5.13) with $x_1/|x|$ and relation (5.14) with $x_2/|x|$, respectively, we obtain, after adding them up
$$\varphi\rho\left(v_{0,111}\frac{x_1}{|x|} + v_{0,112}\frac{x_2}{|x|} + v_{0,122}\frac{x_1}{|x|} + v_{0,222}\frac{x_2}{|x|}\right) = R_A^2 f'(v_0)\mathcal{E}(\rho).$$

The desired result follows since $\mathcal{W}(\rho)$ is the solution of (5.11). □

Lemma 5.1.3 *Given $\rho \in h^{4+\alpha}(\mathbb{S})$, we have that*

$$\partial_\nu(\mathcal{E}(\rho)) = \alpha_A \rho, \qquad (5.15)$$

where $\alpha_A := \partial^2 U/\partial r^2(1, R_A^2) > 0$ and U is the solution of (3.24).

Proof From the definition of $\mathcal{E}(\rho)$, we obtain that the gradient of $\mathcal{E}(\rho)$ is given by the relations

$$\frac{\partial \mathcal{E}(\rho)}{\partial x_1}(x) = \left(v_{0,11} \frac{x_1}{|x|} + v_{0,12} \frac{x_2}{|x|} + v_{0,1} \frac{x_2^2}{|x|^3} - v_{0,2} \frac{x_1 x_2}{|x|^3} \right) \rho\varphi$$

$$- \frac{x_2}{|x|^2} \left(v_{0,1} \frac{x_1}{|x|} + v_{0,2} \frac{x_2}{|x|} \right) \rho'\varphi + \frac{x_1}{|x|} \left(v_{0,1} \frac{x_1}{|x|} + v_{0,2} \frac{x_2}{|x|} \right) \rho\varphi',$$

$$\frac{\partial \mathcal{E}(\rho)}{\partial x_2}(x) = \left(v_{0,12} \frac{x_1}{|x|} + v_{0,22} \frac{x_2}{|x|} - v_{0,1} \frac{x_1 x_2}{|x|^3} + v_{0,2} \frac{x_1^2}{|x|^3} \right) \rho\varphi$$

$$+ \frac{x_1}{|x|^2} \left(v_{0,1} \frac{x_1}{|x|} + v_{0,2} \frac{x_2}{|x|} \right) \rho'\varphi + \frac{x_2}{|x|} \left(v_{0,1} \frac{x_1}{|x|} + v_{0,2} \frac{x_2}{|x|} \right) \rho\varphi',$$

for all $x \neq 0$. Here $\varphi = \varphi(|x| - 1)$, and for x near the unit circle $\mathbb{S}$ we have that $\varphi \equiv 1$. Hence,

$$x_1 \frac{\partial \mathcal{E}(\rho)(x)}{\partial x_1} + x_2 \frac{\partial \mathcal{E}(\rho)(x)}{\partial x_2} = (v_{0,11} x_1^2 + 2 v_{0,12} x_1 x_2 + v_{0,22} x_2^2) \rho(x) x \qquad (5.16)$$

for all $x \in \mathbb{S}$. From Theorem 3.1.15, we also know that $v_0(x) = U(|x|, R_A^2)$ for all $x \in \Omega$. Therefore, $v_0(r\cos(s), r\sin(s)) = U(r, R_A^2)$ for all $s \in \mathbb{R}$ and $0 \leq r \leq 1$. Differentiating this expression two times with respect to the variable r we obtain at $r = 1$

$$v_{0,11} x_1^2 + 2 v_{0,12} x_1 x_2 + v_{0,22} x_2^2 = \frac{\partial^2 U}{\partial r^2}(1, R_A^2) = \alpha_A \quad \text{on} \quad \mathbb{S},$$

and, in view of (5.16), we are done if we show that $\alpha_A > 0$.

Recall that the function U is non-decreasing in the first variable. Combining (3.24) and (3.32) we have that

$$\frac{\partial^2 U}{\partial r^2}(1, R_A^2) = R_A^2 f(U(1, R_A^2)) - \frac{\partial U}{\partial r}(1, R_A^2)$$

$$= R_A^2 f(U(1, R_A^2)) - R_A^2 \int_0^1 rf\left(U(r, R_A^2)\right) dr$$

$$> R_A^2 f(U(1, R_A^2)) - R_A^2 \int_0^1 f\left(U(1, R_A^2)\right) dr \geq 0.$$

This completes the proof. $\square$

Further on, we prove that the normal derivative $\partial_\nu \mathcal{W}$ is a Fourier multiplier operator and we determine its symbol. For this reasoning, we consider expansions of the form

$$\begin{aligned} \mathcal{W}(rx) &= \sum_{k \in \mathbb{Z}} A_k(r) x^k \\ \rho(x) &= \sum_{k \in \mathbb{Z}} \widehat{\rho}(k) x^k \end{aligned} \quad (5.17)$$

with $r \in [0,1]$ and $x \in \mathbb{S}$. Substituting these expressions into the first two equations of (5.10), and comparing coefficients of x^k, we come to the following problems for the unknown functions A_k

$$\begin{cases} A_k'' + \frac{1}{r} A_k' - \frac{k^2}{r^2} A_k = R_A^2 f'(v_0) A_k, & 0 < r < 1, \\ A_k(1) = -v_0'(1) \widehat{\rho}(k). \end{cases} \quad (5.18)$$

We have used here the relation $\partial_\nu v_0 = v_0'(1)$ on $\mathbb{S}$, where we identify again v_0 with its restriction to the interval $[0,1]$.

In order to prove the existence and uniqueness of the solution to (5.18) we consider first the associated problem (5.19). The solution of (5.18) will be then expressed in terms of the solution of this new system.

Lemma 5.1.4 *Given $n \in \mathbb{N}$, the problem*

$$\begin{cases} u'' + \frac{2n+1}{r} u' = R_A^2 f'(v_0) u, & 0 < r < 1, \\ u(0) = 1, \\ u'(0) = 0, \end{cases} \quad (5.19)$$

has a unique solution $u_n \in C^2([0,1])$.

Proof Multiplying the first equation of problem (5.19) with r^{2n+1} and then integrating yields that every solution of (5.19) solves the integral equation

$$u(r) = 1 + \int_0^r \frac{R_A^2}{s^{2n+1}} \int_0^s \tau^{2n+1} f'(v_0(\tau)) u(\tau) \, d\tau \, ds, \quad 0 \le r \le 1. \quad (5.20)$$

We shall use the Banach fixed point theorem to prove that this integral equation possesses a unique solution defined on a small interval $[0, h]$, $h > 0$. To this scope, we obtain from relation (5.20) the following estimation

$$|u(r) - 1| \le R_A^2 \int_0^r \int_0^s f'(v_0(\tau)) |u(\tau)| \, d\tau \, ds$$

$$\le M \frac{r^2}{2} \sup_{[0,1]} |u(\tau)|,$$

where $M := R_A^2 \max_{[0,1]} f'(v_0)$. We introduce now the Banach space

$$\mathbb{E} = \{u : [0, h] \to \mathbb{R} \, : \, u \text{ continuous}\},$$

whereby $h := \min\{1/\sqrt{M}, 1\}$. Furthermore, let $T : \overline{D}_{\mathbb{E}}(1, 1) \to \overline{D}_{\mathbb{E}}(1, 1)$ be the operator defined by

$$(Tu)(r) = 1 + R_A^2 \int_0^r \frac{1}{s^{2n+1}} \int_0^s \tau^{2n+1} f'(v_0(\tau)) u(\tau) \, d\tau \, ds, \quad 0 \le r \le h.$$

From the estimate above we find that T is a well-defined operator. Furthermore, T is a $1/2$–contraction. Indeed, given $u, v \in \overline{D}_{\mathbb{E}}(1, 1)$, we have

$$|Tu - Tv|(r) \le \int_0^r \int_0^s M \sup_{[0,1]} |u(\tau) - v(\tau)| \, d\tau \, ds$$

$$\le \frac{h^2}{2} M \, ||u - v||_\infty \le \frac{1}{2} ||u - v||_\infty$$

for all $0 \le r \le h$. We infer from the Banach fixed point theorem that T has a unique fixed point $u \in \overline{D}_{\mathbb{E}}(1, 1)$, meaning that the equation (5.20) has a unique solution in $u_n \in C([0, h])$. We still have to show that this solution belongs to $C^2([0, h])$. Obviously, u_n is smooth on $(0, h]$. Given $r > 0$, we have that

$$u'_n(r) = \frac{R_A^2 \int_0^r \tau^{2n+1} f'(v_0(\tau)) u_n(\tau) \, d\tau}{r^{2n+1}},$$

and by l'Hospital's theorem we get $\lim_{t\to 0} u'(t) = 0$. Differentiating the relation above once more, we obtain

$$u_n''(r) = -(2n+1)\frac{u_n'(r)}{r} + R_A^2 f'(v_0) u_n(r)$$

for all $0 < r \le h$. Letting $r \searrow 0$, it follows

$$\lim_{r\to 0} \frac{u_n'(r)}{r} = \lim_{r\to 0} \frac{R_A^2 \int_0^r s^{2n+1} f'(v_0(s)) u_n(s)\, ds}{r^{2n+2}}$$

$$\stackrel{l'H}{=} R_A^2 \lim_{t\to 0} \frac{f'(v_0(r)) u_n(r)}{2n+2} = R_A^2 \frac{f'(v_0(0)) u_n(0)}{2n+2} = R_A^2 \frac{f'(c)}{2n+2},$$

where $c \in (0,1)$ (see (3.10)). Consequently, $u_n \in C^2([0,h])$, and

$$u_n''(0) = -(2n+1) R_A^2 \frac{f'(c)}{2n+2} + R_A^2 f'(c) \underbrace{u(0)}_{=1} = R_A^2 \frac{f'(c)}{2n+2}.$$

Since the first equation in (5.19) is linear in u and has no singularities different from 0, the solution found on the interval $[0, h]$ may be extended to the whole interval $[0, 1]$. Notice also that, $v_n(x) = u_n(|x|)$, $x \in \Omega$ solves $\Delta v_n = R_A^2/(2n+1) f'(v_0) v_n$, therefore $u_n \in C^\infty([0,1])$. The uniqueness of u_n follows from (5.20) by making use of Gronwall's lemma. $\square$

We solve now the system (5.18). The existence of solutions to (5.18) is an immediate consequence of Lemma 5.1.4 and the uniqueness follows by a contradiction argument.

Lemma 5.1.5 *Given $k \in \mathbb{Z}$, problem (5.18) possesses a unique solution $A_k \in C^2([0,1])$. Moreover, $A_k \in C^\infty([0,1])$ and*

$$A_k(r) = \frac{-v_0'(1)}{u_{|k|}(1)} \widehat{\rho}(k) r^{|k|} u_{|k|}(r), \quad 0 \le r \le 1. \tag{5.21}$$

We denoted by $u_n, n \in \mathbb{N}$, the solution of (5.19).

Proof We verify first that the function A_k given by (5.21) is a solution of (5.18). The smoothness of this solution follows directly from the smoothness of the solutions of (5.19). Differentiating (5.21) with respect to the variable r, we obtain

$$A'_k(r) = \frac{-v'_0(1)}{u_{|k|}(1)}\widehat{\rho}(k)r^{|k|}u'_{|k|}(r) + \frac{-v'_0(1)}{u_{|k|}(1)}\widehat{\rho}(k)|k|r^{|k|-1}u_{|k|}(r),$$

$$A''_k(r) = \frac{-v'_0(1)}{u_{|k|}(1)}\widehat{\rho}(k)\left[r^{|k|}u''(r) + 2|k|r^{|k|-1}u'_{|k|}(r) + |k|(|k|-1)r^{|k|-2}u_{|k|}(r)\right]$$

for all $r \in [0,1]$. Further on, we get

$$A''_k(r) + \frac{1}{r}A'_k(r) - \frac{k^2}{r^2}A_k(r) =$$

$$= \frac{-v'_0(1)}{u_{|k|}(1)}\widehat{\rho}(k)\left[r^{|k|}u''_{|k|}(r) + 2|k|r^{|k|-1}u'_{|k|}(r) + |k|(|k|-1)r^{|k|-2}u_{|k|}(r)\right.$$

$$\left. + |k|r^{|k|-2}u_{|k|}(r) + r^{|k|-1}u'_{|k|}(r) - k^2 r^{|k|-2}u_{|k|}(r)\right]$$

$$= \frac{-v'_0(1)}{u_{|k|}(1)}\widehat{\rho}(k)r^{|k|}\underbrace{\left[u''_{|k|}(r) + \frac{2|k|+1}{r}u'_{|k|}(r)\right]}_{R_A^2 f'(v_0)u_{|k|}(r)}$$

$$= R_A^2 f'(v_0)A_k.$$

Moreover, $A_k(1) = -v_0(1)\widehat{\rho}(k)$, and we are left to prove the uniqueness within the class $C^2([0,1])$.

Assume by contradiction that we found two solutions u and v of (5.18) for the same $k \in \mathbb{Z}$. Setting $w := u - v$, we have that

$$\begin{cases} w'' + \frac{1}{r}w' - \frac{k^2}{r^2}w = R_A^2 f'(v_0)w, & 0 < r < 1, \\ w(1) = 0. \end{cases} \quad (5.22)$$

We prove now that $\omega(0) = 0$. Indeed, if $\omega(0) \neq 0$ we get from the first equation of the system above, that

$$r^2 w'' + rw' = k^2 w + R_A^2 r^2 f'(v_0)w,$$

which in turn implies
$$(r^2\omega')' = r\omega' + k^2\omega + R_A^2 r^2 f'(v_0)\omega.$$

Integrating this equation with respect to r, we come to the following relation
$$r^2\omega' = \int_0^r \left[s\omega'(s) + k^2\omega(s) + R_A^2 s^2 f'(v_0(s))\omega\right] ds$$

Letting $r \searrow 0$, it follows that
$$\lim_{r\to 0}\omega'(r) = \lim_{r\to 0}\frac{\int_0^r [s\omega'(s) + k^2\omega(s) + R_A^2 s^2 f'(v_0(s))\omega] \, ds}{r^2}$$
$$= \lim_{r\to 0}\frac{r\omega'(r) + k^2\omega(r) + R_A^2 f'(v_0(r))\omega(r)}{2r}$$
$$\xrightarrow[r\to 0]{} \begin{cases} +\infty, & \text{if } \omega(r) > 0, \\ -\infty, & \text{if } \omega(r) < 0, \end{cases}$$

in contradiction with $\omega \in C^2([0,1])$. Assuming that w (or $-w$) has a positive maximum, this value must be take in an interior point of the interval $[0, 1]$, which is not possible. This completes the proof. $\square$

From Lemma 5.1.5 we get the expansion (5.17) of $\mathcal{W}(\rho)$
$$\mathcal{W}(\rho)(rx^k) = \sum_{k\in\mathbb{Z}}\left[-\frac{v_0'(1)}{u_{|k|}(1)}r^{|k|}u_{|k|}(r)\right]\widehat{\rho}(k)x^k \qquad (5.23)$$

for all $\rho \in h^{4+\alpha}(\mathbb{S})$. Moreover, from the first equation of (3.24) we find that the constant α_A, determined in Lemma 5.1.3, satisfies the relation $\alpha_A = R_A^2 f(1) - v_0'(1)$. Moreover, from the same equation and (3.32) we have that
$$v_0'(1) = AR_A^2/2, \qquad (5.24)$$

hence
$$\alpha_A = R_A^2\left(f(1) - \frac{A}{2}\right). \qquad (5.25)$$

With these preparations we give now the main result of this section:

Theorem 5.1.6 *Given* $\rho = \sum_{k \in \mathbb{Z}} \widehat{\rho}(k) x^k \in h^{4+\alpha}(\mathbb{S})$, *we have that*

$$\partial \Phi(0) \left[\sum_{k \in \mathbb{Z}} \widehat{\rho}(k) x^k \right] = \sum_{k \in \mathbb{Z}} \mu_k \widehat{\rho}(k) x^k, \qquad (5.26)$$

where the symbol $(\mu_k)_{k \in \mathbb{Z}}$ *is given by the relation*

$$\mu_k := -\frac{1}{R_A^3}|k|^3 + \frac{1}{R_A^3}|k| - G\left(\frac{A}{2}\frac{u'_{|k|}(1)}{u_{|k|}(1)} + A - f(1)\right) \qquad (5.27)$$

for $k \in \mathbb{Z}$. *Moreover,*

$$\sigma(\partial \Phi(0)) = \{\mu_k \; : \; k \in \mathbb{Z}\}. \qquad (5.28)$$

Proof The proof follows straightforward from relations (4.22), (5.23) and Lemmas 5.1.2 and 5.1.3. □

5.2 Estimates for the symbol of the derivative of Φ

In order to study the stability properties of the unique radially symmetric equilibrium determined in Chapter 3 we need to study the sign of symbol μ_k, $k \in \mathbb{Z}$, of the Fourier multiplier $\partial \Phi(0)$ in dependence of $(A, G) \in (0, f(1)) \times (0, \infty)$. We consider in here just the case when $G > 0$, since for $G < 0$ we already established in Theorem 3.0.3 (d) that this circular equilibrium is unstable. The situation when $G = 0$ is discussed in Section 5.4. It is worth noticing that $\mu_k = \mu_{-k}$ for all $k \in \mathbb{Z}$, so that we consider just the terms μ_k with $k \in \mathbb{N}$.

At first we have to ascertain that 0 is in the spectrum of $\partial \Phi(0)$ for all $(A, G) \in (0, f(1)) \times [0, \infty)$.

Proposition 5.2.1 *We have that*

$$\mu_1 = 0.$$

Proof We noticed in the proof of Lemma 5.1.3 that $v_0 = \mathcal{T}(0)$ satisfies $v_0(r) = U(r, R_A^2)$ for all $0 \leq r \leq 1$, thus

$$v_0'' + \frac{1}{r}v_0' = R_A^2 f(v_0) \quad \text{in } (0, 1).$$

Differentiating this equation with respect to r and setting $w_0 := v_0'$ we find out that
$$\begin{cases} w_0'' + \frac{1}{r}w_0' - \frac{1}{r^2}w_0 = R_A^2 f'(v_0)w_0, \\ w_0(1) = v_0'(1). \end{cases} \quad (5.29)$$

Given $r \in [0,1]$ we define $w(r) = v_0'(1)ru_1(r)/u_1(1)$, where u_1 denotes the solution of the problem (5.19) when $n = 1$. With $c := v_0'(1)/u_1(1)$, we obtain by differentiation that
$$w'(r) = c(u_1(r) + ru_1'(r)),$$
$$w''(r) = c(ru_1''(r) + 2u_1'(r)),$$
which in turn implies that w is a solution of the equation
$$w'' + \frac{1}{r}w' - \frac{1}{r^2}w = c(ru_1''(r) + 3u_1'(r)) = R_A^2 f'(v_0)w.$$

Moreover, for $r = 1$, we get
$$w(1) = v_0'(1)\frac{u_1(1)}{u_1(1)} = v_0'(1).$$

It can be seen, as in the proof of Lemma 5.1.5, that the solution of (5.29) is unique, hence $w_0 = w$, meaning that $v_0'(r) = v_0'(1)ru_1(r)/u_1(1)$ for all $0 \leq r \leq 1$. Rearranging this relation it follows that
$$\frac{v_0'(r)}{v_0'(1)r} = \frac{u_1(r)}{u_1(1)},$$
which leads to
$$\frac{u_1'(r)}{u_1(1)} = \frac{v_0''(r)\, r\, v_0'(1) - v_0'(r)\, v_0'(1)}{v_0'(1)^2\, r^2} = \frac{1}{v_0'(1)}\left[\frac{v_0''(r)}{r} - \frac{v_0'(r)}{r^2}\right].$$

In $r = 1$ we obtain
$$\frac{u_1'(1)}{u_1(1)} = \frac{1}{v_0'(1)}[v_0''(1) - v_0'(1)] = \frac{v_0''(1)}{v_0'(1)} - 1.$$

In view of relations (5.24) and (5.25), and the definition of α_A, we have
$$\frac{u_1'(1)}{u_1(1)} = \frac{-\frac{AR_A^2}{2} + R_A^2 f(1)}{\frac{AR_A^2}{2}} - 1 = -2 + \frac{2f(1)}{A}.$$

Inserting this result in the expression of μ_1 yields

$$\mu_1 = -\frac{1}{R_A^3} + \frac{1}{R_A^3} - G\left[\frac{A}{2}\left(-2 + \frac{2f(1)}{A}\right) + A - f(1)\right] = 0.$$

□

We show now that the sequence $\mu_k \to_{k\to\infty} -\infty$. This is obviously true if the sequence $(u'_n(1)/u_n(1))_{n\in\mathbb{N}}$ is bounded. Even more, we prove:

Lemma 5.2.2 *It holds that*
$$\frac{u'_n(1)}{u_n(1)} \xrightarrow[n\to\infty]{} 0,$$

Proof Let $n \in \mathbb{N}$ and set $v := u_{n+1} - u_n \in C^2([0,1])$. Recall that u_n is an increasing function for all $n \in \mathbb{N}$. This can be seen by differentiating relation (5.20). From (5.19) we obtain

$$\begin{cases} v'' + \dfrac{2n+1}{r}v' = R^2 f'(v_0)v - \dfrac{2}{r}u'_{n+1}, & 0 < r < 1, \\ v'(0) = 0, \\ v(0) = 0. \end{cases}$$

Furthermore, we obtain

$$\lim_{t\to 0}\frac{v'(t)}{t} = \lim_{t\to 0}\frac{u'_{n+1}(t) - u'_n(t)}{t} = R_A^2 f'(c)\left(\frac{1}{2n+4} - \frac{1}{2n+2}\right)$$

$$= -R_A^2 f'(c)\frac{2}{(2n+4)(2n+2)} < 0,$$

which implies that $v'(t) < 0$ for $t \in (0,\delta)$ and some $\delta < 1$. Thus, v is decreasing on $(0,\delta)$. Let now $m_t := \min_{[0,t]} v \le 0$ for all $t \in [0,1]$. It must hold that $m_t = v(t) \le 0$, which implies $u_{n+1}(t) \le u_n(t)$ for all $t \in [0,1]$. Particularly, $u_{n+1}(1) \le u_n(1)$. It also holds that

$$u'_n(t) = R_A^2 \int_0^t \left(\frac{\tau}{t}\right)^{2n+1} f'(v_0(\tau))u_n(\tau)\,d\tau$$

$$\ge R_A^2 \int_0^t \left(\frac{\tau}{t}\right)^{2n+3} f'(v_0(\tau))u_{n+1}(\tau)\,d\tau$$

$$= u'_{n+1}(t)$$

for all $t \in [0,1]$, hence $u'_{n+1}(1) \leq u'_n(1)$. Moreover, from

$$u'_n(t) = \frac{R_A^2 \int_0^t \tau^{2n+1} f'(v_0(\tau)) u_n(\tau)\, d\tau}{t^{2n+1}}$$

it follows that

$$u'_n(1) = R_A^2 \int_0^1 \tau^{2n+1} f'(v_0(\tau)) u_n(\tau)\, d\tau.$$

The Dominated convergence theorem implies $u'_n(1) \searrow 0$. A similar argument provides $u_n(1) \searrow 1$. Therefore, $u'_n(1)/u_n(1) \to 0$, and we are done. $\square$

We consider now more closely the term of μ_k, $k \in \mathbb{N}$. Given $k \in \mathbb{N}$ large enough, we find in view of the previous lemma a unique value G_k of the parameter G such that $\mu_k = 0$ iff $G = G_k$. The next lemma asserts that the sequence $(G_k)_{k \geq k_1}$ is strictly increasing, provided $k_1 \in \mathbb{N}$ is far away from 0. This lemma will play an important role in the proof of Theorem 6.0.6.

Lemma 5.2.3 *Given $k \in \mathbb{N}$ such that*

$$\frac{A}{2}\frac{u'_k(1)}{u_k(1)} + A - f(1) \neq 0,$$

we set

$$G_k := \frac{-\frac{1}{R^3}k^3 + \frac{1}{R^3}k}{\frac{A}{2}\frac{u'_k(1)}{u_k(1)} + A - f(1)}. \tag{5.30}$$

There exists $k_1 \in \mathbb{N}$ with the property that $0 < G_k < G_{k+1}$ for all $k \geq k_1$.

Proof From Lemma 5.2.2 we find $k_0 \in \mathbb{N}$ such that the quotient of the ratio (5.30) is negative for all $k \geq k_0$. From Proposition 5.2.1 we deduce that $k_0 \geq 2$. Given $k \in \mathbb{N}$, we set $a_k := u'_k(1)$ and $b_k := u_k(1) - 1$. From Lemma 5.2.2 we have that $a_k \searrow 0$ and $b_k \searrow 0$. We prove first some estimates for the sequences (a_k) and (b_k) in the sense that, we find a positive constant $M > 0$ such that

$$a_k \leq \frac{M}{2k+2} \quad \text{and} \quad b_k \leq \frac{M}{(2k+1)(2k+3)} \tag{5.31}$$

for all $k \in \mathbb{N}$. Indeed, with $M := R_A^2 u_0(1) \max_{[0,1]} f'(v_0)$ we obtain, due to (5.20), that

$$b_k = \int_0^1 R_A^2 \int_0^s \tau^{2k+1} f'(v_0(\tau)) u_k(\tau) \, d\tau$$

$$\leq M \int_0^1 \int_0^s \tau^{2k+1} \, d\tau = \frac{M}{(2k+1)(2k+3)}.$$

The estimate for (a_k) follows similarly. Furthermore, by partial integration we come to

$$k \cdot (a_k - a_{k+1}) = R_A^2 \int_0^1 k\tau^{2k+1} f'(v_0(\tau)) [u_k(\tau) - \tau^2 u_{k+1}(\tau)] \, d\tau$$

$$= R_A^2 k \frac{\tau^{2k+2}}{2k+2} f'(v_0(\tau)) [u_k(\tau) - \tau^2 u_{k+1}(\tau)] \Big|_0^1$$

$$- \int_0^1 k \frac{\tau^{2k+2}}{2k+2} \left\{ f''(v_0(\tau)) \left[u_k(\tau) - \tau^2 u_{k+1}(\tau) \right] \right.$$

$$\left. + f'(v_0(\tau)) \left[u_k'(\tau) - 2\tau u_{k+1}(\tau) - \tau^2 u_{k+1}'(\tau) \right] \right\} d\tau$$

$$\leq R_A^2 f'(1) \frac{k}{2k+2} (u_k(1) - u_{k+1}(1)) + L \frac{k}{(k+1)^2},$$

with a constant L independent of k. Letting now $k \to \infty$ we get

$$k \cdot (a_k - a_{k+1}) \xrightarrow[k \to \infty]{} 0. \tag{5.32}$$

Having this estimates at hand we can prove now the assertion of the lemma. For simplicity we set $C := 2(f(1) - A)/A > 0$. For $k \geq k_0$, we have that

$$G_k = \frac{2}{AR^3} \frac{k^3 - k}{C - \dfrac{u_k'(1)}{u_k(1)}}.$$

Further on, we compute

$$G_{k+1} > G_k \Leftrightarrow \frac{(k+1)^3 - (k+1)}{C - \frac{a_{k+1}}{1+b_{k+1}}} > \frac{k^3 - k}{C - \frac{a_k}{1+b_k}}$$

$$\Leftrightarrow C(k+1)^3 - C(k+1) - Ck^3 + Ck$$
$$> \frac{a_{k+1}}{1+b_{k+1}}(k - k^3) + \frac{a_k}{1+b_k}[(k+1)^3 - (k+1)]$$

$$\Leftrightarrow C(3k^2 + 3k) > \frac{a_{k+1}(k - k^3) + a_k(k^3 + 3k^2 + 2k)}{(1+b_k)(1+b_{k+1})}$$
$$+ \frac{a_{k+1}b_k(k - k^3) + a_k b_{k+1}(k^3 + 3k^2 + 2k)}{(1+b_k)(1+b_{k+1})}$$

$$\Leftrightarrow C(3k^2 + 3k) > \frac{k^3(a_k - a_{k+1}) + a_{k+1}k + a_k(3k^2 + 2k)}{(1+b_k)(1+b_{k+1})}$$
$$+ \frac{a_{k+1}b_k(k - k^3) + a_k b_{k+1}(k^3 + 3k^2 + 2k)}{(1+b_k)(1+b_{k+1})}.$$

Taking into consideration the relations (5.31) and (5.32), we find a positive integer $k_1 \geq k_0$ such that strict inequality holds in the last relation above for all $k \geq k_1$. $\square$

5.3 Instability for $G > G_*$

We prove now that if G is large enough, then the radially symmetric equilibrium determined in Theorem 3.0.2 is unstable, in the sense that problem (4.20) has a nontrivial backwards solution. This reveals that the exponential stability result stated in Theorem 3.0.3 (a) gives a false impression about the stability properties of the radially symmetric equilibrium.

To prove the statement, we set

$$G_* := \min \left\{ G_k : \frac{A}{2} \frac{u_k'(1)}{u_k(1)} + A - f(1) < 0 \right\}. \qquad (5.33)$$

Lemma 5.2.3 yields that $0 \leq G_*$. Moreover, since $G_k \to_{k\to\infty} \infty$, the minimum must be achieved, i.e. we find at least a number $k_0 \in \mathbb{N}$ such that

$$\frac{A}{2}\frac{u'_{k_0}(1)}{u_{k_0}(1)} + A - f(1) < 0, \tag{5.34}$$

and $G_* = G_{k_0}$. With these preparations we state:

Theorem 5.3.1 *Let G_* be the constant defined by (5.33). The radially symmetric equilibrium determined in Theorem 3.0.2 is unstable for all $G > G_*$.*

Proof Let $G > G_*$ be given and k_0 be an integer such that $G_* = G_{k_0}$ and (5.34) holds true. It follows that

$$-\frac{1}{R^3}k_0^3 + \frac{1}{R^3}k_0 > G\left(\frac{A}{2}\frac{u'_{k_0}(1)}{u_{k_0}(1)} + A - f(1)\right),$$

which implies $\mu_{k_0} > 0$. We are left to check the following instability assumptions

$$\begin{cases} \sigma_+(\partial\Phi(0)) = \sigma(\partial\Phi(0)) \cap \{\lambda \in \mathbb{C} : \text{Re}\lambda > 0\} \neq \emptyset, \\ \inf\{\text{Re}\lambda : \lambda \in \sigma_+(\partial\Phi(0))\} > 0, \end{cases}$$

where $\partial\Phi(0)$ stands here for the complexification of $\partial\Phi(0)$. The first one is clear since $\mu_{k_0} > 0$. Moreover, we infer from Lemma 5.2.2 that $\mu_k \to_{k\to\infty} -\infty$, and therefore the unstable spectrum $\sigma_+(\partial\Phi(0))$ is a finite set of positive eigenvalues of the Fréchet derivative $\partial\Phi(0)$. Thus, we found out that the assumptions of [45, Theorem 9.1.3] are satisfied and therewith the radially symmetric solution in unstable. □

5.4 Exponential stability for G = 0

As we mentioned before, when $G = 0$ the problem (2.1) is exactly the Hele-Shaw problem in a horizontal cell, cf. [29, 31]. Determining the equilibria of problem (2.1) for $G = 0$ is equivalent to solving the free boundary problem

$$\begin{cases} \Delta p = 0 & \text{in } \Omega_\rho, \\ p = \kappa_\rho & \text{on } \Gamma_\rho, \\ \partial_{\nu_\rho} p = 0 & \text{on } \Gamma_\rho. \end{cases} \tag{5.35}$$

Assuming that $\rho \in \mathcal{V}$ is known, we obtain from the first and the third equation of (5.35) that p must be a constant function. Consequently, the curvature of Γ_ρ is constant and Γ_ρ must be a circle. Furthermore, it is obvious that any circle determines a steady-state solution for (2.1). Hence, all the equilibria of (2.1) are circles. For simplicity, we study therefore the stability properties of the equilibrium when the tumor occupies the unitary disc. That is why we take the constant R, fixed at the beginning of Chapter 4, to be $R = 1$, meaning that functions in $\mathcal{V}$ parametrise curves near the unit circle.

We show in the main result of this section, Theorem 5.4.3, that the stationary solution near $\rho = 0$ build a three dimensional centre manifold which attracts at an exponential rate solutions starting nearby. To prove that result we have to do first some preparation.

From the relations (5.26) and (5.27), we get, for $G = 0$ and $R_A = 1$, that $\mu_k = |k|(1-k^2)$, for $k \in \mathbb{Z}$, hence the derivative of Φ in 0 is the Fourier multiplier

$$\sum_{k \in \mathbb{Z}} \widehat{\rho}(k) x^k \longmapsto \sum_{k \in \mathbb{Z}} |k|(1-k^2) \widehat{\rho}(k) x^k.$$

For simplicity we let $A := \partial \Phi(0)$. The eigenspace corresponding to the eigenvalue $\mu_1 = 0$ is three dimensional spanned by $\{1, x, x^{-1}\}$. Moreover, there exists also an open neighbourhood $\mathcal{U}$ of the origin in $\mathbb{R}^3$ with the property that

$$\mathcal{E} := \{(\rho_{(c,R)}, 1/(R+1)) \in \mathcal{V} \times buc^{2+\alpha}(\Omega) : (c, R) \in \mathcal{U}\}$$

contains all the equilibria of problem (4.20) in $\mathcal{V}$. Given $(c, R) \in \mathcal{U}$, the mapping

$$\rho_{(c,R)}(x) = \sqrt{(R+1)^2 - |c|^2 + <c, x>^2} + <c, x> - 1, \quad x \in \mathbb{S},$$

defines an element of $\mathcal{V}$ and the boundary $\Gamma_{\rho_{(c,R)}}$ is exactly the circle with centre c and radius $(R+1)$. In order to study the stability of the 0 solution for the problem (4.20) we shall proceed like in [25, 31, 45] and construct a three dimensional invariant submanifold of the phase space with the property that the eigenspace corresponding to the eigenvalue $\mu_1 = 0$ is tangential to it in 0.

Let $0 < \beta < \alpha < 1$ and $\theta := (\alpha - \beta)/3$. Letting $\mathcal{G}(\rho) := \Phi(\rho) - A\rho$ for $\rho \in \widetilde{\mathcal{V}} := \{\rho \in h^{4+\beta}(\mathbb{S}) : \|\rho\|_{C(\mathbb{S})} < 1/4\}$, we write problem (4.20) as follows

$$\begin{aligned} \partial_t \rho &= A\rho + \mathcal{G}(\rho), \quad t \geq 0, \\ \rho(0) &= \rho_0, \end{aligned} \quad (5.36)$$

where $-A \in \mathcal{H}(h^{4+\beta}(\mathbb{S}), h^{1+\beta}(\mathbb{S}))$. We have $\mathcal{G} \in C^\infty(\mathcal{V}, D_A(\theta))$, $\mathcal{G}(0) = 0$ as well as $\partial \mathcal{G}(0) = 0$. These relations follow by reason of

$$D_A(\theta) = (h^{4+\beta}(\mathbb{S}), h^{1+\beta}(\mathbb{S}))_\theta = h^{1+\alpha}(\mathbb{S})$$

and $D_A(\theta + 1) = \{\rho \in h^{4+\beta}(\mathbb{S}) : A\rho \in D_A(\theta)\} = h^{4+\alpha}(\mathbb{S})$. Notice that the part of A in $D_A(\theta)$ is denoted again by A. We refer also to [45] for details. Let us denote by $P \in \mathcal{L}(h^{1+\beta}(\mathbb{S}))$ the spectral projection associated with the nonnegative spectral set $\{0\}$

$$P = \frac{1}{2\pi i} \int_C R(z, A) \, dz,$$

where C is the circle centred in 0 with radius $1/2$. Notice that the closed ball bounded by C contains non of the negative eigenvalues of the operator A. Using the Fourier expansions of the functions $\rho \in h^{1+\beta}(\mathbb{S})$ we see that P is a Fourier multiplier. More precisely, given $\rho = \sum_{k \in \mathbb{Z}} \widehat{\rho}(k) x^k \in C^\infty(\mathbb{S})$ we compute

$$\begin{aligned}
P\left[\sum_{k \in \mathbb{Z}} \widehat{\rho}(k) x^k\right] &= \frac{1}{2\pi i} \int_C R(z, A) \rho \, dz = \frac{1}{2\pi i} \int_C \sum_{k \in \mathbb{Z}} \frac{1}{z - \mu_k} \widehat{\rho}(k) x^k \, dz = \\
&= \frac{1}{2\pi i} \sum_{k \in \mathbb{Z}} \int_C \frac{1}{z - \mu_k} \widehat{\rho}(k) x^k \, dz = \\
&= \sum_{|k| \leq 1} \left(\frac{1}{2\pi i} \int_C \frac{1}{z - \mu_k} \, dz\right) \widehat{\rho}(k) x^k = \\
&= \sum_{|k| \leq 1} \widehat{\rho}(k) x^k.
\end{aligned}$$

Notice that for $\rho \in h^{1+\alpha}(\mathbb{S})$ the series $\sum_{k \in \mathbb{Z}} [1/(z - \mu_k)] \widehat{\rho}(k) x^k$ is uniformly convergent. Particularly, Theorem 4.5.1 implies $P \in \mathcal{L}(h^{k+\sigma}(\mathbb{S}))$ for all $k \in \mathbb{N}$ and $\sigma \in (0, 1)$.

We set $X_1 = P(h^{1+\beta}(\mathbb{S}))$, $X_2 = (I - P)(h^{1+\beta}(\mathbb{S}))$ and

$$A_1 : X_1 \to X_1, \quad A_1 x = Ax$$

$$A_2 : X_2 \cap h^{4+\beta}(\mathbb{S}) \to X_2, \quad A_2 x = Ax.$$

Then $A_1 \in L(X_1)$ is the 0_{X_1} operator, $\sigma(A_2) = \{\mu_k : k \geq 2\}$ and $D_{A_2}(\theta) = D_A(\theta) \cap X_2$, $D_{A_2}(\theta + 1) = D_A(\theta + 1) \cap X_2$. Setting

$$\widetilde{h}^{k+\sigma}(\mathbb{S}) := \{\rho \in h^{k+\sigma}(\mathbb{S}) : \widehat{\rho}(m) = 0 \quad \text{for} \quad |m| \leq 1\}$$

for $k \in \mathbb{N}$ and $\sigma \in (0,1)$, we have
$$D_{A_2}(\theta) = \widetilde{h}^{1+\alpha}(\mathbb{S}) \quad \text{and} \quad D_{A_2}(\theta+1) = \widetilde{h}^{4+\alpha}(\mathbb{S}).$$

Moreover, A_2 is the part in $\widetilde{h}^{1+\alpha}(\mathbb{S})$ of the operator $-A_2 \in \mathcal{H}(\widetilde{h}^{4+\beta}(\mathbb{S}), \widetilde{h}^{1+\beta}(\mathbb{S}))$. Let us further notice that $Ph^{k+\sigma}(\mathbb{S})$ is a three-dimensional space, thus
$$h^{k+\sigma}(\mathbb{S}) = Ph^{k+\sigma}(\mathbb{S}) \oplus \widetilde{h}^{k+\sigma}(\mathbb{S}), \ k \in \mathbb{N}, \ \sigma \in (0,1)$$

is a topological direct sum.

Choose $r_0 > 0$ such that $\overline{B}_{h^{4+\alpha}(\mathbb{S})}(0, r_0(1+K)) \subset \mathcal{O}$ and let $\psi : X_1 \to [0,1]$ be a smooth cutoff function satisfying
$$\psi(x) = 1 \quad \text{for} \quad \|x\|_{C^{1+\beta}(\mathbb{S})} \leq \frac{1}{2} \quad \text{and} \quad \psi(x) = 0 \quad \text{for} \quad \|x\|_{C^{1+\beta}(\mathbb{S})} \geq 1.$$

We have denoted by K the norm of the embedding $(X_1, \|\cdot\|_{C^{4+\alpha}(\mathbb{S})}) \hookrightarrow (X_1, \|\cdot\|_{C^{1+\beta}(\mathbb{S})})$. Given $r \leq r_0$, the mapping
$$\mathcal{G}_r : X_1 \times B_{\widetilde{h}^{4+\alpha}(\mathbb{S})}(0, r_0) \subset h^{4+\alpha}(\mathbb{S}) \to h^{1+\alpha}(\mathbb{S}),$$
$$\mathcal{G}_r(\rho) := \mathcal{G}\left(\psi\left(\frac{P\rho}{r}\right)P\rho + (I-P)\rho\right), \quad \rho \in X_1 \times B_{\widetilde{h}^{4+\alpha}(\mathbb{S})}(0, r_0),$$

has the same regularity properties as $\mathcal{G}$ and $\mathcal{G}(0) = 0, \partial \mathcal{G}(0) = 0$. Further on, for $r \leq r_0$ the abstract Cauchy problem
$$\begin{aligned} \partial_t \rho &= A\rho + \mathcal{G}_r(\rho), \quad t \geq 0, \\ \rho(0) &= \rho_0, \end{aligned} \tag{5.37}$$

is equivalent to the problem (5.36) for small solutions. More precisely, the solutions of (5.36) remaining in $B_{X_1}(0, r/2) \times B_{\widetilde{h}^{4+\alpha}(\mathbb{S})}(0, r_0) \subset h^{4+\alpha}(\mathbb{S})$ coincide with the solutions of (5.37). Given $\rho \in X_1 \times B_{\widetilde{h}^{4+\alpha}(\mathbb{S})}(0, r_0)$, we assume that $A + \partial \mathcal{G}_r(\rho)$ is the part in $h^{1+\alpha}(\mathbb{S})$ of a sectorial operator $B : h^{4+\beta}(\mathbb{S}) \subset h^{1+\beta}(\mathbb{S}) \to h^{1+\beta}(\mathbb{S})$, such that $D_B(\theta) = h^{1+\alpha}(\mathbb{S})$ and $D_B(\theta+1) = h^{4+\alpha}(\mathbb{S})$. This can be achieved by choosing r_0 small enough.

Under this assumptions we obtain for each initial data $\rho_0 \in X_1 \times B_{\widetilde{h}^{4+\alpha}(\mathbb{S})}(0, r_0)$, applying again [45, Theorem 8.4.1], existence and uniqueness of a maximal defined solution $\rho \in C([0, T(\rho_0)), X_1 \times B_{\widetilde{h}^{4+\alpha}(\mathbb{S})}(0, r_0)) \cap C^1([0, T(\rho_0)), h^{1+\alpha}(\mathbb{S}))$.

Clearly, problem (5.37) is equivalent to the following coupled system
$$x' = A_1 x + f(x, y),$$
$$y' = A_2 y + g(x, y), \qquad (5.38)$$
$$x(0) = x_0, \quad y(0) = y_0,$$

where
$$f : X_1 \times B_{\widetilde{h}^{4+\alpha}(\mathbb{S})}(0, r_0) \subset X_1 \times \widetilde{h}^{4+\alpha}(\mathbb{S}) \to X_1$$
is given by
$$f(x, y) = P\mathcal{G}\left(\psi\left(\frac{x}{r}\right)x + y\right),$$
and
$$g : X_1 \times B_{\widetilde{h}^{4+\alpha}(\mathbb{S})}(0, r_0) \subset X_1 \times \widetilde{h}^{4+\alpha}(\mathbb{S}) \to \widetilde{h}^{1+\alpha}(\mathbb{S})$$
is defined by
$$g(x, y) = (I - P)\mathcal{G}\left(\psi\left(\frac{x}{r}\right)x + y\right).$$
Being interested in the stability of the equilibria to problem (4.20) located near the trivial solution 0 it will be sufficient to consider the problem (5.38) for small r. A pair (x, y) is called solution of (5.38) if there exists a constant $T > 0$ such that
$$x \in C^1([0, T], Ph^{1+\alpha}(\mathbb{S})),$$
$$y \in C^1([0, T], \widetilde{h}^{1+\alpha}(\mathbb{S})) \cap C([0, T], \widetilde{h}^{4+\alpha}(\mathbb{S})),$$
and if (x, y) satisfies the system (5.38) pointwise. In view of [45, Proposition 9.2.1] we can choose r_0 small enough to guarantee, for $r \leq r_0$, the existence of a constant positive $C(r)$ with the property that solutions to (5.38) satisfying initially $\|\rho_0\|_{C^{4+\alpha}(\mathbb{S})} \leq C(r)$ exist globally.

We state now a theorem on the existence and smoothness of invariant manifolds for the system (5.38), result which can be found in [49, Theorem 4.1] and [45, Theorem 9.2.2] (see also [25] and [31]).

Theorem 5.4.1 (Existence of centre manifolds) *Given* $k \in \mathbb{N}_{>0}$, *there exists* $r_k \in (0, r_0]$ *and for each* $r \in (0, r_k]$ *there is a unique mapping*
$$\sigma \in BC^k(X_1, \widetilde{h}^{4+\alpha}(\mathbb{S})),$$
satisfying
$$\sigma(0) = 0 \qquad \partial \sigma(0) = 0.$$

Moreover
$$\|\sigma(x) - \sigma(\overline{x})\|_{C^{4+\alpha}(\mathbb{S})} \leq b\|x - \overline{x}\|_{C^{1+\beta}(\mathbb{S})}$$
for a suitable constant b and
$$\|\sigma(x)\|_{C^{4+\alpha}(\mathbb{S})} \leq r, \quad \forall x \in X_1.$$

Let $\mathcal{M} := \mathcal{M}(r, k) = \{(x, \sigma(x)) : x \in X_1\} \subset h^{4+\alpha}(\mathbb{S})$. Then $\mathcal{M}$ is a globally invariant 3-dimensional C^k-manifold for the problem (5.38), i.e. given $(x_0, y_0) \in \mathcal{M}$, the solution (x, y) to (5.38) exists in the large and $(x(t), y(t)) \in \mathcal{M}$ for $t \geq 0$.

Denote by $z(\,\cdot\,) = z(\,\cdot\,, x, \sigma)$ the global solution of the initial value problem for the reduced ordinary differential equation

$$\begin{aligned} z'(t) &= f(z(t), \sigma(z(t))), \quad t \in \mathbb{R}, \\ z(0) &= x. \end{aligned} \qquad (5.39)$$

The function σ is the unique fixed point of the following equation

$$\sigma(x) = \int_{-\infty}^{0} e^{-tA_2} g(z(t, x, \sigma), \sigma(z(t, x, \sigma)))\, dt, \qquad (5.40)$$

and for $(x_0, y_0) \in \mathcal{M}$ we have that $(x(t), y(t)) = (z(t, x_0, \sigma), \sigma(z(t, x_0, \sigma)))$, $t \geq 0$ is the globally defined solution to (5.38).

Additionally, if $\rho : \mathbb{R} \to h^{4+\alpha}(\mathbb{S})$ is a globally defined solution of (4.20) with

$$\rho(t) \in W(r) := B_{X_1}\left(0, \frac{r}{2}\right) \times B_{\overline{h}^{4+\alpha}(\mathbb{S})}(0, r),$$

i.e. $\|P\rho(t)\|_{C^{1+\beta}(\mathbb{S})} < r/2$ and $\|(I - P)\rho(t)\|_{C^{4+\alpha}(\mathbb{S})} < r$ for all $t \geq 0$, then $(I - P)\rho(t) = \sigma(P\rho(t))$ and $P\rho$ is the unique solution of the following initial value problem

$$\begin{aligned} z'(t) &= f(z(t), \sigma(z(t))), \quad t \in \mathbb{R}, \\ z(0) &= P\rho_0. \end{aligned}$$

Thus, $\mathcal{M}$ contains all small global solutions of (4.20). The tangent space to $\mathcal{M}$ in 0 is X_1, the eigenspace corresponding to the eigenvalue 0

$$T_0(\mathcal{M}) = im(\mathrm{id}_{X_1}, \partial\sigma(0)) = X_1 \times \{0\} \cong X_1.$$

Fix now $k \geq 2$. Given $r \in (0, r_k]$, we construct now a locally invariant C^k–manifold $\mathcal{M}_{loc}^c$ for the problem (4.20) containing just stationary solutions of (4.20). Let $V \subset \mathcal{U}$ be a small neighbourhood of 0 in $\mathbb{R}^3$ satisfying

$$\rho_{(c,R)} \in W(r), \forall (c,R) \in V.$$

By Theorem 5.4.1 we have then $\rho_{(c,R)} = (P\rho_{(c,R)}, \sigma(P\rho_{(c,R)}))$ for $(c, R) \in V$. The mapping $[\mathcal{U} \ni (c, R) \mapsto \rho(c, R) := \rho_{(c,R)} \in h^{4+\alpha}(\mathbb{S})]$ is smooth and we compute

$$\partial \rho(0,0)[c, R] = \frac{c}{2}x^{-1} + R + \frac{\bar{c}}{2}x = \langle c, x \rangle + R, \quad [c, R] \in \mathbb{R}^3. \quad (5.41)$$

Given $(c, R) \in \mathcal{U}$, we can represent the periodic function $P\rho_{(c,R)}$ uniquely by its trigonometric series

$$P\rho_{(c,R)} = \widetilde{R} + \langle \widetilde{c}, x \rangle, \quad (5.42)$$

where $\widetilde{R} = \widehat{\rho}_{(c,R)}(0)$ and $\widetilde{c} = 2\widehat{\rho}_{(c,R)}(-1)$.

Using this relation, we state that the mapping $\mathcal{F} : \mathcal{U} \to \mathbb{R}^3$, defined by $\mathcal{F}(c, R) := (\widetilde{c}, \widetilde{R})$, where $(\widetilde{c}, \widetilde{R})$ are given by (5.42), is smooth, satisfies $F(0) = 0$ and additionally, by (5.42), $\partial \mathcal{F}(0) = id_{\mathbb{R}^3}$. If V is small enough, then $\mathcal{F} : V \to \mathcal{F}(V)$ is a smooth diffeomorphism. Given $(\widetilde{c}, \widetilde{R}) \in \mathcal{F}(V)$ we have

$$P\rho\mathcal{F}^{-1}(\widetilde{c}, \widetilde{R}) = \widetilde{R} + \langle \widetilde{c}, x \rangle,$$

thus $P\rho\mathcal{F}^{-1}$ is the restriction to $\mathcal{F}(V)$ of the isomorphism $[\mathbb{R}^3 \ni (\widetilde{c}, \widetilde{R}) \mapsto \langle \widetilde{c}, x \rangle + \widetilde{R} \in X_1]$. We conclude that $P\rho(V)$ is an open neighbourhood of 0 in X_1. Define $\mathcal{M}_{loc}^c$ as the graph of the restriction of σ to the open set $P\rho(V)$. We have obtain in this way a local invariant manifold for the system (4.20). The example of A. Kelley, see [56, (Example 13.7)], shows that invariant manifolds are in general not unique. In the context of our problem we know additionally

$$\mathcal{M}_{loc}^c = \{(x, \sigma(x)) : x \in P\rho(V)\} = \{(P\rho_{(c,R)}, \sigma(P\rho_{(c,R)})) : (c, R) \in V\} =$$
$$= \{\rho_{(c,R)} : (c, R) \in V\}.$$

This means that the (a priori non-unique) invariant manifold $\mathcal{M}_c^{loc}$ consists in equilibria only, i.e. in circles, and is therefore unique.

The manifold $\mathcal{M}$ attracts the solutions of (5.38) for small initial data. This result is found in [45]. More precisely we have:

Theorem 5.4.2 *Given $\omega \in (0, -\lambda_2 = 6)$, there exist positive constants $M = M(\omega)$ and $\bar{r} = \bar{r}(\omega)$ such that for all $r \leq \bar{r}$, $(x_0, y_0) \in X_1 \times \widetilde{h}^{4+\alpha}(\mathbb{S})$ with $\|(x_0, y_0)\|_{C^{4+\alpha}(\mathbb{S})} \leq C(r)$, the solution (x, y) to (5.38) exists in the large and satisfies*

$$\|y(t) - \sigma(x(t))\|_{C^{4+\alpha}(\mathbb{S})} \leq M e^{-\omega t} \|y_0 - \sigma(x_0)\|_{C^{4+\alpha}(\mathbb{S})} \quad \text{for} \quad t \in [0, \infty).$$

The following result on asymptotic stability ensures that for small initial data ρ_0 the solution to (4.20) exists in the large and there exists a steady state belonging to the local centre manifold $\mathcal{M}_{loc}^c$, uniquely determined by the initial data, which attracts the solution exponentially. The proof of this result follows similarly as the proof of [31, Theorem 6.5], which is an adoption of the proof of [49, Proposition 9.2.4].

Theorem 5.4.3 *Let $\omega \in (0, 6)$ and $r \leq \bar{r}$ be given. There exist positive constants $K = K(\omega)$ and a neighbourhood $\mathcal{V}(r)$ of 0 in $h^{4+\alpha}(\mathbb{S})$ such that, for $\rho_0 \in \mathcal{V}(r)$ the solution to (4.20) exists in the large and there exist $z_0 \in P\rho(V)$ such that*

$$\|\rho(t) - (z_0, \sigma(z_0))\|_{C^{4+\alpha}(\mathbb{S})} \leq K e^{-\omega t} \|(I - P)\rho_0 - \sigma(P\rho_0))\|_{C^{4+\alpha}(\mathbb{S})} \quad \text{for } t \in [0, \infty).$$

Notice that for $z_0 \in P\rho(V)$, the mapping $(z_0, \sigma(z_0))$ belongs to the local centre manifold $\mathcal{M}_{loc}^c$ and is uniquely determined by the initial data ρ_0.

5.5 Exponential convergence of $2\pi/l$–periodic data

In this section we show that the exponential stability result stated in Theorem 3.0.3 (a) can be generalised for solutions of the original problem (2.1) which correspond to initial data ρ_0 closed to the unique radially symmetric equilibrium and $2\pi/l$–periodic. The positive integer l depends on the constant G, which is in this section positive, so that $D(0, R_A)$ is the unique equilibrium of the problem (2.1). The main result of this section, Theorem 5.5.1 requires the following assumption

$$\frac{A}{2} \frac{u_0'(1)}{u_0(1)} + A - f(1) > 0, \tag{5.43}$$

which means that $\mu_0 = \mu_0(G) < 0$ for all $G > 0$. This assumption is satisfied for example if $R_A = 1$ and $f = \text{id}_{[0, \infty)}$, cf. Observation 6.0.7. By restricting problem (2.1) to appropriate subspaces of the small Hölder spaces, we obtain the following stability result:

Theorem 5.5.1 *Assume that (5.43) holds true. Given $G > 0$, there exists a positive integer $l_G \in \mathbb{N}$ such that for all $\omega \in (0, \mu_0)$ and $l \geq l_G$ we find positive constants $K_l > 0$ and $\delta_l > 0$ with the property that if $\|\rho_0\|_{C^{4+\alpha}(\mathbb{S})} \leq \delta_l$ and ρ_0 is $2\pi/l$–periodic, then the solution ρ to (4.20) exists in the large, and*

$$\|\rho(t)\|_{C^{4+\alpha}(\mathbb{S})} + \|\rho'(t)\|_{C^{1+\alpha}(\mathbb{S})} \leq K_l e^{-\omega t} \|\rho_0\|_{C^{4+\alpha}(\mathbb{S})}, \quad t \geq 0.$$

Moreover, the solution ρ is $2\pi/l$–periodic for all $t \geq 0$.

In order to prove this theorem we introduce first appropriate subspaces of the small Hölder spaces. Given $k \in \mathbb{N}$ and $l \in \mathbb{N}, n \geq 2$, we define the subspace of $h^{k+\alpha}(\mathbb{S})$ consisting of $2\pi/l$–periodic functions by

$$h_l^{k+\alpha}(\mathbb{S}) := \{\rho \in h^{k+\alpha}(\mathbb{S}) \,:\, \rho(x) = \rho(e^{2\pi i/l} x) \quad \text{for all } x \in \mathbb{S}\}.$$

Set further $\mathcal{V}_l := \mathcal{V} \cap h_l^{4+\alpha}(\mathbb{S})$. The Fourier series associated to $\rho \in h_l^{k+\alpha}(\mathbb{S})$ is

$$\rho = \sum_{k=0}^{\infty} \widehat{\rho}(kl) x^{kl},$$

where $\widehat{\rho}(kl)$ is the kl–th Fourier coefficient of ρ. Indeed, since $\rho(x) = \rho(xe^{2\pi i/l})$ for all $x \in \mathbb{S}$, we deduce by identifying the Fourier expansions of ρ and $\rho(\cdot e^{2\pi i/l})$ that

$$\sum_{k \in \mathbb{Z}} \widehat{\rho}(k) x^k = \sum_{k \in \mathbb{Z}} \widehat{\rho}(k) e^{2k\pi i/l} x^k,$$

hence, if $\widehat{\rho}(k) \neq 0$, then k is a multiple of l, and the formula is proved.

Our goal is to prove that if $\rho \in \mathcal{V}_l$, then $\Phi(\rho) \in h_l^{1+\alpha}(\mathbb{S})$, that is

$$\Phi \in C^{\infty}(\mathcal{V}_l, h_l^{1+\alpha}(\mathbb{S})). \tag{5.44}$$

Having shown (5.44), by choosing l large enough we can exclude the eigenvalues μ_k with k small from the spectrum of $\partial \Phi(0)$. These are the eigenvalues which we could not estimate whether they are negative or not. In this way we also eliminate $\mu_1 = 0$ from the spectrum.

Let $\rho \in \mathcal{V}_l$ be given and let $\psi := \Theta_*^\rho v$, where v is the solution of (4.18). We prove that ψ satisfies $\psi(x) = \psi(e^{2\pi i/l} x)$ for all $x \in \Omega_\rho$. Therefore, we must prove first that, if $x \in \Omega_\rho$, then $xe^{2\pi i/l}$ belongs to Ω_ρ. Indeed, given $x \in \Omega_\rho$, we have that

$$|xe^{2\pi i/l}| = |x| < R\left(1 + \rho\left(\frac{x}{|x|}\right)\right) = R\left(1 + \rho\left(\frac{xe^{2\pi i/l}}{|xe^{2\pi i/l}|}\right)\right),$$

which implies that $xe^{2\pi i/l} \in \Omega_\rho$. The function $\psi \in buc^{2+\alpha}(\Omega_\rho)$ is the unique solution of the Dirichlet problem

$$\begin{cases} \Delta\psi = f(\psi) & \text{in } \Omega_\rho, \\ \psi = 1 & \text{on } \Gamma_\rho. \end{cases} \qquad (5.45)$$

We define $\overline{\psi}(x) := \psi(xe^{2\pi i/l})$. We have defined in this way a further solution of (5.45), since

$$\Delta\overline{\psi}(x) = \Delta\psi(xe^{2\pi i/l}) = f(\psi(xe^{2\pi i/l})) = f(\overline{\psi}(x)),$$

for all $x \in \Omega_\rho$ and $\overline{\psi} = 1$ on Γ_ρ. The uniqueness of the solution to (5.45) implies that $\psi = \overline{\psi}$. Thus $\psi(x) = \psi(xe^{2\pi i/l})$ for all $x \in \Omega_\rho$.

Following the same schema we can prove that $p := \Theta_*^\rho q$, where q is the solution of (4.19), satisfies $p(x) = p(e^{2\pi i/l}x)$ for all $x \in \Omega_\rho$ provided $\rho \in \mathcal{V}_l$. With these preparations we state:

Lemma 5.5.2 *Given $l \geq 2$, the operator Φ maps smoothly $\mathcal{V}_l$ into $h_l^{1+\alpha}(\mathbb{S})$.*

Proof We have to show that $\mathcal{B}_1(\rho, \mathcal{T}(\rho))$ and $\mathcal{B}_1(\rho, \mathcal{S}(\rho))$ are $2\pi/l$–periodic. The assertion follows then in view of (4.21) and (4.26). We prove the assertion just for $\mathcal{B}_1(\rho, \mathcal{T}(\rho))$, the proof that $\mathcal{B}_1(\rho, \mathcal{S}(\rho))$ is $2\pi/l$–periodic follows analogously. Indeed, given $x \in \mathbb{S}$ we have

$$\mathcal{B}_1(\rho, \mathcal{T}(\rho))(x) = \langle \nabla(\Theta_*^\rho v)(\Theta_\rho(x)), \nabla N_\rho(\Theta_\rho(x)) \rangle$$

$$= \left\langle \nabla\psi(\Theta_\rho(x)), \frac{\nabla N_\rho(\theta_\rho(x))}{|\nabla N_\rho(\Theta_\rho(x))|} \right\rangle |\nabla N_\rho(\Theta_\rho(x))|$$

$$= \partial_\nu \psi(\Theta_\rho(x)) |\nabla N_\rho(\Theta_\rho(x))|,$$

where ψ is the solution of (5.45). From relation (4.23) we obtain that

$$|\nabla N_\rho (\Theta_\rho (xe^{2\pi i/l}))| = |\nabla N_\rho(\Theta_\rho(x))|$$

for all $x \in \mathbb{S}$. In order to prove that also $\partial_\nu \psi(\Theta_\rho(xe^{2\pi i/l})) = \partial_\nu \psi(\Theta_\rho(x))$, $x \in \mathbb{S}$, we introduce the rotation matrix

$$M = \begin{pmatrix} \cos\left(\frac{2\pi}{l}\right) & -\sin\left(\frac{2\pi}{l}\right) \\ \sin\left(\frac{2\pi}{l}\right) & \cos\left(\frac{2\pi}{l}\right) \end{pmatrix}.$$

We infer, from $\psi(x) = \psi\left(xe^{2\pi i/l}\right) = \psi(M \cdot x)$, that $\nabla\psi(x) = M^\top \cdot \nabla\psi(M \cdot x)$ for all $x \in \Omega_\rho$. Thus,

$$\partial_\nu \psi \left(\Theta_\rho \left(xe^{2\pi i/l}\right)\right) = \langle \nabla\psi \left(\Theta_\rho \left(M \cdot x\right)\right), \nu_\rho \left(\Theta_\rho \left(M \cdot x\right)\right) \rangle$$

$$= \left\langle \nabla\psi \left(M \cdot \Theta_\rho(x)\right), M \cdot \frac{x - \rho'(x)(-x_2, x_1)}{\sqrt{(1+\rho)^2 + \rho'^2(x)}} \right\rangle$$

$$= \left\langle M \cdot \nabla\psi(\Theta_\rho(x)), M \cdot \frac{x - \rho'(x)(-x_2, x_1)}{\sqrt{(1+\rho)^2 + \rho'^2(x)}} \right\rangle$$

$$= \left\langle \nabla\psi(\Theta_\rho(x)), M^\top \cdot M \cdot \frac{x - \rho'(x)(-x_2, x_1)}{\sqrt{(1+\rho)^2 + \rho'^2(x)}} \right\rangle$$

$$= \langle \nabla\psi \left(\Theta_\rho(x)\right), \nu_\rho \left(\Theta_\rho(x)\right) \rangle$$

$$= \partial_\nu \psi \left(\Theta_\rho(x)\right)$$

for all $x \in \mathbb{S}$. Summarising, $\mathcal{B}_1(\rho, \mathcal{T}(\rho))\left(xe^{2\pi i/l}\right) = \mathcal{B}_1(\rho, \mathcal{T}(\rho))(x)$ for all $x \in \mathbb{S}$, and the proof is completed.

$\square$

The Fréchet derivative of the mapping $\Phi \in C^\infty(\mathcal{V}_l, h_l^{1+\alpha}(\mathbb{S}))$ is, in view of (5.26), given by the relation

$$\partial\Phi(0)\left[\sum_{k\in\mathbb{Z}} \widehat{\rho}(kl)x^{kl}\right] = \sum_{k\in\mathbb{Z}} \mu_{kl}\widehat{\rho}(kl)x^{kl},$$

where $(\mu_{kl})_k$ are defined by (5.27). We come now to the proof of the main result of this chapter:

Proof *(Proof of Theorem 5.5.1)* Let $G > 0$ be given. Since $\mu_{|k|} \to_{|k|\to\infty} -\infty$, we find a positive integer l_G such that $\mu_{|k|} \leq \mu_0$ for all $|k| \geq l_G$. Let $l \geq l_G$ be fixed. In view of relation Lemma 5.5.2, we find that the restriction $\Phi \in C^\infty(\mathcal{V}_l, h_l^{1+\alpha}(\mathbb{S}))$ satisfies the assumptions of [45, Theorem 9.1.2].

Indeed, since $l \geq l_G$, we find that $\mu_{kl} \leq \mu_0$ for all $k \in \mathbb{N}$. That the spectrum of the complexification of the Fréchet derivative of the restriction $\Phi \in$

$C^\infty(\mathcal{V}_l, h_l^{1+\alpha}(\mathbb{S}))$ in 0 consists just of the eigenvalues $\{\mu_{kl} \ : \ k \in \mathbb{N}\}$ can be seen easily, by using Theorem 4.5.1.

Hence, the spectrum of $\partial\Phi(0) \in \mathcal{L}(h_l^{4+\alpha}(\mathbb{S},\mathbb{C}), h_l^{1+\alpha}(\mathbb{S},\mathbb{C}))$ is bounded away from the positive half plane by μ_0. The assertion follows now immediately from [45, Theorem 9.1.1]. □

Chapter 6

Local bifurcation analysis

In this chapter we take again $(A, G) \in (0, f(1)) \times (0, \infty)$. As we noticed in Section 5, the radially symmetric equilibrium is unstable if G is sufficiently large. The question we follow in this section is whether this property is not related to the existence of other equilibria than the trivial, radially symmetric state $D(0, R_A)$.

We answer this question by considering the operator Φ, defined by relation (4.21), as an operator depending on both variables $(G, \rho) \in \mathbb{R} \times \mathcal{V}$. The regularity results presented in Section 3 imply that $\Phi \in C^\infty(\mathbb{R} \times \mathcal{V}, h^{1+\alpha}(\mathbb{S}))$. Determining the equilibria of (4.21), reduces to finding the solutions of the operator equation

$$\Phi(G, \rho) = 0. \tag{6.1}$$

The main tool we use is the classical bifurcation result for bifurcations from simple eigenvalues due to Crandall and Rabinowitz:

Theorem 6.0.3 (see [13, 18]) *Let X, Y be real Banach spaces and $G(\lambda, u)$ be a C^q ($q \geq 3$) mapping from a neighbourhood of a point $(\lambda_0, u_0) \in \mathbb{R} \times X$ into Y. Let the following assumptions hold:*

(i) $G(\lambda_0, u_0) = 0$, $\partial_\lambda G(\lambda_0, u_0) = 0$,

(ii) $\operatorname{Ker} \partial_u G(\lambda_0, u_0)$ *is one dimensional, spanned by* v_0,

(iii) $\operatorname{Im} \partial_u G(\lambda_0, u_0)$ *has codimension 1*,

(iv) $\partial_\lambda \partial_\lambda G(\lambda_0, u_0) \in \operatorname{Im} \partial_u G(\lambda_0, u_0)$, $\partial_\lambda \partial_u G(\lambda_0, u_0) v_0 \notin \operatorname{Im} \partial_u G(\lambda_0, u_0)$.

Then (λ_0, u_0) is a bifurcation point of the equation

$$G(\lambda, u) = 0 \qquad (6.2)$$

in the following sense: In a neighbourhood of (λ_0, u_0) the set of solutions of equation (6.2) consists of two C^{q-2} curves Σ_1 and Σ_2, which intersect only at the point (λ_0, u_0). Furthermore, Σ_1, Σ_2 can be parameterised as follows:

$\Sigma_1 : (\lambda, u(\lambda)), |\lambda - \lambda_0|$ *is small*, $u(\lambda_0) = u_0, u'(\lambda_0) = 0,$

$\Sigma_2 : (\lambda(\varepsilon), u(\varepsilon)), |\varepsilon|$ *is small*, $(\lambda(0), u(0)) = (\lambda_0, u_0), u'(0) = v_0.$

In here, the branch of trivial solutions is the set $\Sigma := \{(G, 0) : G \in \mathbb{R}\}$, and the Fréchet derivative of Φ with respect to ρ at $(G, 0)$ is, cf. (5.26), a Fourier multiplier with a symbol which depends smoothly on G, i.e.

$$\partial_\rho \Phi(G, 0)[\rho] = \sum_{k \in \mathbb{Z}} \mu_k(G) \widehat{\rho}(k) x^k \qquad (6.3)$$

for all $\rho = \sum_{k \in \mathbb{Z}} \widehat{\rho}(k) x^k \in h^{4+\alpha}(\mathbb{S})$, where $(\mu_k(G))_{k \in \mathbb{Z}}$ are given by the relation (5.27). As we seen in Lemma 5.2.1, $\mu_1(G) = 0$ for all $G \in \mathbb{R}$. Moreover, the transversality condition (second condition in (iv)) is not satisfied, since the mixed second derivative

$$\partial_G \partial_\rho \Phi(0) \left[\sum_{k \in \mathbb{Z}} \widehat{\rho}(k) x^k \right] = -\sum_{k \in \mathbb{Z}} \left(\frac{A}{2} \frac{u'_{|k|}(1)}{u_{|k|}(1)} + A - f(1) \right) \widehat{\rho}(k) x^k, \qquad (6.4)$$

maps x, the eigenvalue of $\partial_\rho \Phi(G, 0)$ corresponding to the eigenvalue $\mu_1(G)$, into the image of $\partial_\rho \Phi(G, 0)$, i.e. $\partial_G \partial_\rho \Phi(G, 0)[x] = 0 \in \text{Im} \, \partial_\rho \Phi(G, 0)$. This makes a direct application of Theorem 6.2 in this setting impossible.

Hence, we have to eliminate somehow the eigenvalue $\mu_1(G)$ from the spectrum of $\partial \Phi(G, 0)$, and also to reduce the dimensions of the eigenspace corresponding to an eigenvalue $\mu_k(G)$, which we may chose to be equal to 0 if G is large enough, to one. This is due to the fact that the dimension of the eigenspace corresponding to an arbitrary eigenvalue $\mu_k(G)$ is larger then 2, since x^k and x^{-k} are eigenvectors of this eigenvalue. We have already determined a method to eliminate eigenvalues. In Lemma 5.5.2 we restricted the operator Φ to spaces consisting of $2\pi/l$−periodic functions and eliminated in this manner the set $\{\mu_k : 1 \leq |k| \leq l - 1\}$ from the spectrum of complexification of the Fréchet derivative of Φ. So, by choosing l large we can reduce the dimension of the kernel of

$\partial\Phi(G,0)$ to be 2. Moreover, considering only spaces of even functions, we find that the kernel of the restriction of $\partial\Phi(G,0)$ is a one-dimensional subspace.

We proceed and introduce subspaces of the small Hölder spaces consisting only of $2\pi/l-$periodic and even functions. Given $k \in \mathbb{N}$ and $l \in \mathbb{N}, l \geq 2$, we define

$$h_{e,l}^{k+\alpha}(\mathbb{S}) := \{\rho \in h^{k+\alpha}(\mathbb{S}) : \rho(x) = \rho(\overline{x}) \text{ and } \rho(x) = \rho(e^{2\pi i/l}x) \quad \forall x \in \mathbb{S}\},$$

where for $x \in \mathbb{C}$, $\overline{x}$ denotes the complex conjugate of x. Set further $\mathcal{V}_{e,l} := \mathcal{V} \cap h_{e,l}^{4+\alpha}(\mathbb{S})$. By identifying functions on $\mathbb{S}$ with $2\pi-$periodic functions on $\mathbb{R}$ we can expand $\rho \in h_{e,l}^{k+\alpha}(\mathbb{S})$ in the following way

$$\rho(s) = \sum_{k=0}^{\infty} a_k \cos(kls),$$

where $a_k = 2\widehat{\rho}(kl)$ for $k \geq 0$. Indeed, since $h_{e,l}^{k+\alpha}(\mathbb{S})$ is a subspace of $h_l^{k+\alpha}(\mathbb{S})$, we have that

$$\rho = \sum_{k \in \mathbb{Z}} \widehat{\rho}(kl) x^{kl}$$

for all $\rho \in h_{e,l}^{k+\alpha}(\mathbb{S})$. Moreover, since ρ is even

$$a_k = a_{-k} = 2\widehat{\rho}(kl)$$

for all $k \in \mathbb{Z}$. With $x = e^{is}$ we obtain

$$\rho = \sum_{k \in \mathbb{Z}} \widehat{\rho}(kl) x^{kl} = \sum_{k \in \mathbb{Z}} \frac{a_k}{2} e^{ikls} = \sum_{k=0}^{\infty} a_k \cos(kls).$$

With this preparation we state:

Lemma 6.0.4 *Given $l \geq 2$, the operator Φ maps smoothly $\mathbb{R} \times \mathcal{V}_{e,l}$ into $h_{e,l}^{1+\alpha}(\mathbb{S})$, i.e.*

$$\phi \in C^{\infty}(\mathbb{R} \times \mathcal{V}_{e,l}, h_{e,l}^{1+\alpha}(\mathbb{S})).$$

Proof The proof is similar to that of Lemma 5.5.2 and therefore we omit it.

$\square$

Fix now $l \geq 2$. We infer from relation (6.3) that the Fréchet derivative of the smooth mapping $\Phi : \mathbb{R} \times \mathcal{V}_{e,l} \to h_{e,l}^{1+\alpha}(\mathbb{S})$ at $(G,0)$ is the Fourier multiplier

$$\partial_\rho \Phi(G,0) \left[\sum_{k=0}^{\infty} a_k \cos(kls) \right] = \sum_{k=0}^{\infty} \mu_{kl}(G) a_k \cos(kls) \qquad (6.5)$$

for all $\rho = \sum_{k=0}^{\infty} a_k \cos(klx) \in h_{e,l}^{4+\alpha}(\mathbb{S})$. Notice that the eigenspace generated by $\{x, x^{-1}\}$ is no longer contained in the kernel of $\partial_\rho \Phi(G, 0)$. We give now a last lemma before proving the main result of this chapter, Theorem 6.0.6, which is obtained by applying the theorem on bifurcations from simple eigenvalues to the restriction of Φ defined in Lemma 6.0.4.

Lemma 6.0.5 *Assume that*

$$\frac{A}{2}\frac{u_0'(1)}{u_0(1)} + A - f(1) \neq 0 \tag{6.6}$$

and set

$$G_\bullet := \frac{\frac{1}{R_A^3}k_1^3 - \frac{1}{R_A^3}k_1}{\min_{0 \leq k \leq k_1}\left\{\left|f(1) - A - \frac{A}{2}\frac{u_k'(1)}{u_k(1)}\right| : f(1) - A - \frac{A}{2}\frac{u_k'(1)}{u_k(1)} \neq 0\right\}},$$

with k_1 the positive integer determined in Lemma 5.2.3. Given $G > G_\bullet$ and $l \geq 2$, there exists at most a positive integer k such that $\mu_{kl}(G) = 0$.

Proof Our first assumption (6.6) implies that $\mu_0(G) \neq 0$. Furthermore, we notice that if

$$\frac{A}{2}\frac{u_{kl}'(1)}{u_{kl}(1)} + A - f(1) = 0, \tag{6.7}$$

then $kl \geq 2$, thus $\mu_{kl}(G) \neq 0$. Hence, if $\mu_{kl}(G) = 0$, then equality does not hold in (6.7). Whence, if $\mu_{kl}(G) = 0$, then it must hold that $kl > k_1$.

Indeed, if $\mu_{kl}(G) = 0$ for some $kl \leq k_1$, we obtain from our assumption $G > G_\bullet$ that

$$G_\bullet < G = \frac{\frac{1}{R_A^3}(kl)^3 - \frac{1}{R_A^3}kl}{\left|f(1) - A - \frac{A}{2}\frac{u_{kl}'(1)}{u_{kl}(1)}\right|} \leq G_\bullet.$$

If $kl > k_1$, then $\mu_{kl}(G) = 0$ iff

$$G = \frac{-\frac{1}{R_A^3}(kl)^3 + \frac{1}{R_A^3}kl}{\frac{A}{2}\frac{u_{kl}'(1)}{u_{kl}(1)} + A - f(1)} = G_{kl}.$$

From Lemma 5.2.3 we know that the sequence $(G_{kl})_k$ is increasing from $kl \geq k_1$, whence there exists at most a $k \in \mathbb{N}$ with $G_{kl} = G$, and the proof is completed. $\square$

We come now to the main theorem of this chapter which states that for certain values of the parameter G there exists local bifurcation branches from the trivial, radially symmetric solution which consist only of stationary solutions of the original problem (2.1).

Theorem 6.0.6 *Assume that* (6.6) *holds true. Let $l \geq 2$ be fixed and $k \in \mathbb{N}$, $k \geq 1$ such that $G_{kl} > G_\bullet$. The pair $(G_{kl}, 0)$ is a bifurcation point from the trivial solution Σ. More precisely, in a suitable neighbourhood of $(G_{kl}, 0)$, there exists a smooth branch of solutions $\left(G^{kl}(\varepsilon), \rho^{kl}(\varepsilon)\right)$ of problem* (6.1). *For $\varepsilon \to 0$, we have the following asymptotic expressions:*
$$G^{kl}(\varepsilon) = G_{kl} + O(\varepsilon),$$
$$\rho^{kl}(\varepsilon) = \varepsilon \cos(kls) + O(\varepsilon^2).$$
Moreover, any other $G > G_\bullet$, $G \notin \{G_{kl} : k \geq 1\}$, is not a bifurcation point.

Proof Let $k \in \mathbb{N}$ be given such that $G_{kl} > G_\bullet$, where G_{kl} are the constants defined in Lemma 5.2.3. Lemma 6.0.5 ensures that $\mu_{ml}(G_{kl}) \neq 0$ for $m \neq k$. Our assumption (6.6) ensures that the Fréchet derivative $\partial_\rho \Phi(G_{ml}, 0)$ of the restriction $\Phi : \mathbb{R} \times \mathcal{V}_{e,l} \subset h_{e,l}^{4+\alpha}(\mathbb{S}) \to h_{e,l}^{1+\alpha}(\mathbb{S})$, which is given by relation (6.5), has a one dimensional kernel spanned by $\cos(kls)$. We also observe that its image is closed and has codimension equal to one.

We are left now to prove that the transversality condition (iv) of Theorem 6.0.3 holds. Since $G_{kl} > G_\bullet$, Lemma 6.0.5 ensures that $kl \geq k_1 + 1$ and by Proposition 5.2.3 we find that
$$-\left(\frac{A}{2}\frac{u'_{kl}(1)}{u_{kl}(1)} + A - f(1)\right) > 0.$$
Further on, we get from relation (6.4) that
$$\partial_G \partial_\rho \Phi(0) \left[\cos(kls)\right] = -\left(\frac{A}{2}\frac{u'_{kl}(1)}{u_{kl}(1)} + A - f(1)\right)\cos(kls),$$
and since $\cos(kls) \notin \operatorname{Im} \partial_\rho \Phi(G_{kl}, 0)$ we deduce that the assumption of Theorem 6.0.3 are all verified. By applying Theorem 6.0.3 we obtain the bifurcation result stated in Theorem 6.0.6 and the asymptotic expressions for the bifurcation branches $\left(G^{kl}(\varepsilon), \rho^{kl}(\varepsilon)\right)$.

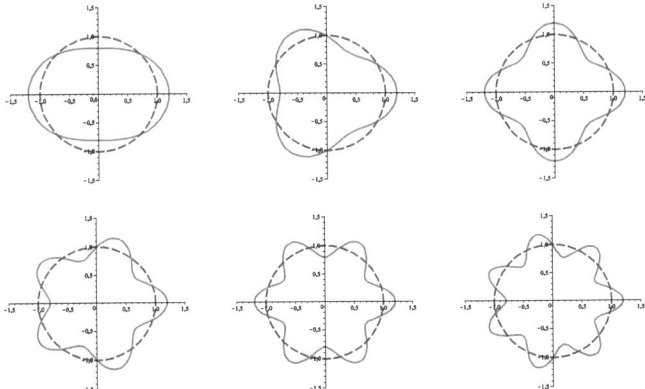

Figure 6.1: Possible steady states of the problem (2.1)

Moreover, if $G > G_\bullet$ and $G \neq G_{kl}$ for all k with $kl \geq k_1 + 1$, then in view of Lemma 6.0.5, $\mu_{kl}(G) \neq 0$ for all $k \in \mathbb{N}$, and we may apply Theorem 4.5.1 to obtain that $\partial_\rho \Phi(G, 0)$ is an isomorphisms. The Implicit function theorem then states that $(G, 0)$ is not a bifurcation point and the proof is completed. $\square$

The possible steady states of problem (2.1) are depicted in Figure 6.1. We present now a situation when (5.43), and particularly (6.6), hold true.

Observation 6.0.7 *The assertion*
$$\frac{A}{2}\frac{u_0'(1)}{u_0(1)} + A - f(1) > 0$$
is fullfield when $f = \mathrm{id}_{[0,\infty)}$ *and* $R_A = 1$.

Proof In view of Proposition 5.2.1, our assertion is equivalent with
$$\frac{A}{2}\frac{u_0'(1)}{u_0(1)} + A - f(1) > 0 = \mu_1(G) = \frac{A}{2}\frac{u_1'(1)}{u_1(1)} + A - f(1).$$

Consequently, we have to show that
$$\frac{u_0'(1)}{u_0(1)} > \frac{u_1'(1)}{u_1(1)}.$$

We assume now that u_0, the solution of (5.19) when $n = 0$, is analytic and the Taylor series associated to u_0 in 0

$$u_0 = \sum_{k=0}^{\infty} a_k x^k,$$

converges on $[0, 1]$. Problem (5.19) writes now as follows

$$\begin{cases} x u_0'' + u_0' - x u_0 = 0, & 0 \leq x \leq 1, \\ u_0(0) = 1, \\ u_0'(0) = 0. \end{cases}$$

From the initial conditions of (5.19) it follows immediately that $a_0 = u_0(0) = 1$ and $a_1 = u_0'(0) = 0$. Plugging u_0 and its derivatives in the first equation of the system, one finds out that

$$\sum_{k=1}^{\infty} k(k+1) a_{k+1} x^k + \sum_{k=0}^{\infty} (k+1) a_{k+1} x^k - a_0 x - \sum_{k=2}^{\infty} a_{k-1} x^k = 0. \qquad (6.8)$$

Identifying the coefficient of x^k in (6.8) yields

$$a_2 = \frac{a_0}{4} = \frac{1}{4},$$

$$a_{k+1} = \frac{a_{k-1}}{(k+1)^2}, \quad \forall k \in \mathbb{N},$$

and, from $a_1 = 0$, we deduce that

$$a_{2k+1} = 0 \quad \text{and} \quad a_{2k} = \prod_{n=1}^{k} \frac{1}{(2n)^2}, \quad \forall k \in \mathbb{N},$$

thus,

$$u_0(x) = 1 + \sum_{k=1}^{\infty} \left(\prod_{n=1}^{k} \frac{1}{(2n)^2} \right) x^{2k}, \quad x \in [0, 1]. \qquad (6.9)$$

We make now the same assumption on u_1, the solution of (5.19) when $n = 1$. We then get, that u_1 is the solution of the following system

$$\begin{cases} xu_1'' + 3u_1' - xu_1 = 0, & 0 \leq x \leq 1, \\ u_1(0) = 1, \\ u_1'(0) = 0. \end{cases}$$

As above, we obtain

$$\sum_{k=1}^{\infty} k(k+1)a_{k+1}x^k + 3\sum_{k=0}^{\infty}(k+1)a_{k+1}x^k - a_0 x - \sum_{k=2}^{\infty} a_{k-1}x^k = 0, \quad (6.10)$$

and therefore

$$a_{2k+1} = 0 \quad \text{and} \quad a_{2k} = \prod_{n=1}^{k} \frac{1}{2n(2n+2)}, \quad \forall k \in \mathbb{N}.$$

Thus,

$$u_1(x) = 1 + \sum_{k=1}^{\infty} \left(\prod_{n=1}^{k} \frac{1}{2n(2n+2)} \right) x^{2k}, \quad x \in [0,1]. \quad (6.11)$$

It is worth noticing that the Taylor series associated to u_0 and u_1, respectively, in 0 define analytic functions on the whole real line, so that the representations (6.9) and (6.11) are valid.

With three exact decimals we have that

$$\frac{u_0'(1)}{u_0(1)} \approx 0.446 > 0.240 \approx \frac{u_1'(1)}{u_1(1)},$$

which leads to the desired conclusion. $\square$

Bibliography

[1] J. A. ADAM: *A mathematical model of tumor growth III. Comparison with experiments*, Math. Biosci. **86**, 213–227 (1987).

[2] T. ALARCÓN, H. M. BYRNE & P. K. MAINI: *A cellular automaton model for tumor growth in inhomogeneous environment*, J. Theor. Biol. **225**, 257–274 (2003).

[3] EDS. J. A. ADAM & N. BELOMO: "A Survey of Models on Tumor Immune System Dynamics", Birkhäuser, 1997.

[4] H. AMANN: "Ordinary Differential Equations. An Introduction to Nonlinear Analysis.", Walter de Gruyter, Berlin, 1990.

[5] H. AMANN: "Linear and Quasilinear Parabolic Problems", Volume I, Birkhäuser, Basel, 1995.

[6] H. AMANN & J. ESCHER: "Analysis III", Birkhäuser, Basel, 2004.

[7] A. R. A. ANDERSON: *A hybrid mathematical model of solid tumor invasion: The importance of cell adhesion*, Math. Med. Biol. **22**, 163–186 (2005).

[8] A. R. A. ANDERSON & M. A. J. CHAPLAIN: *Continuous and discrete mathematical model of tumor-induced angiogenesis*, Bull. Math. Biol. **60**, 905–915 (2006).

[9] W. ARENDT & S. BU: *Operator-valued Fourier multipliers on periodic Besov spaces and applications*, Proceedings of the Edinburgh Mathematical Society, **47**, 15-33 (2004).

[10] N. BELLOMO, N. K. LI & P. K. MAINI: *On the foundations of cancer modelling: Selected topics, speculations, and perspectives* Mathematical Models and Methods in Applied Sciences, **18, No. 4**, 593–647 (2008).

[11] A. BORISOVICH & A. FRIEDMAN: *Symmetric-breaking bifurcation for free boundary problems*, Indiana Univ. Math. J. **54**, 927–947 (2005).

[12] C. J. W. BREWARD, H. M. BYRNE & C. E. LEWIS: *The role of cell-cell interaction in a two-phase model for avascular tumor growth* J. Math. Biol., **45**, 125–152 (2002).

[13] B. BUFFONI & J. TOLAND: "Analytic Theory of Global Bifurcation: An Introduction", Princeton, New Jersey, 2003.

[14] H. M. BYRNE & M. A. CHAPLAIN: *Growth of nonnecrotic tumors in the presence and absence of inhibitors* Math. Biosci., **130**, 151–181 (1995).

[15] H. M. BYRNE, J. R. KING, D. L. S. MCELWAIN & L. PREZIOSI: *A two-phase model of solid tumor growth* Appl. Math. Lett., **16**, 567–573 (2003).

[16] H. M. BYRNE & L. PREZIOSI: *Modelling solid tumor growth using the theory of mixtures* IMA J. Math. Appl. Med. Biol., **20**, 341–366 (2003).

[17] M. A. J. CHAPLAIN: *Avascular growth, angiogensis, and vascular growth in solid tumors: The mathematical modelling of the stages of tumor development*, Math. Comput. Model **23**, 47–87 (1996).

[18] M. G. CRANDALL & P. H. RABINOWITZ: *Bifurcation from simple eigenvalues,* Journal of Functional Analysis ,**8**, 321–340 (1971).

[19] V. CRISTINI, J. LOWENGRUB & Q. NIE: *Nonlinear simulation of tumor growth*, Journal of Mathematical Biology, **46**, 191–224 (2003).

[20] V. CRISTINI, J. LOWENGRUB & Q. NIE: *Nonlinear three-dimensional simulation of solid tumor growth*, Discr. Cont. Dyn. Syst. B, **7**, 581–604 (2007).

[21] S. B. CUI: *Analysis of a free boundary problem modeling tumor growth*, Acta Mathematica Sinica, English Series, **21** (5), 1071–1082 (2005).

[22] S. B. CUI & J. ESCHER: *Bifurcation analysis of an elliptic free boundary problem modelling the growth of avascular tumors*, SIAM J. Math. Anal., **39** (1), 210–235 (2007).

[23] S. B. CUI & A. FRIEDMAN: *Hyperbolic free boundary problem modelling tumor growth*, Interface Free Bound, **5**, 159–181 (2003).

[24] G. DA PRATO & P. GRISVARD: *Equations d'évolution abstraites non-linéaires de type parabolique*, Ann. Mat. Pura Appl., **120**, 329–326 (1979).

[25] G. DA PRATO & A. LUNARDI: *Stability, instability, and center manifold theorem for fully nonlinear autonomous parabolic equations in Banach space*, Arch. Ration. Mech. Anal., **101** (1988), 115–141.

[26] E. DE ANGELIS & L. PREZIOSI: *Advection diffusion models for solid tumors in vivo and related free-boundary problems*, Math. Mod. Meth. Appl. Sci., **10**, 379–408 (2000).

[27] J. ESCHER & A-V. MATIOC: *Radially symmetric growth of nonnecrotic tumors*, to appear in Nonlinear Differential Equations and Applications.

[28] J. ESCHER & B-V. MATIOC: *A moving boundary problem for periodic Stokesian Hele-Shaw flows*, to appear in Interfaces and Free Boundaries.

[29] J. ESCHER, A-V. MATIOC & B.-V. MATIOC: *Classical solutions and stability results for Stokesian Hele-Shaw flows*, to appear in Annali della Scuola Normale Superiore di Pisa, Classe di Scienze.

[30] J. ESCHER & G. SIMONETT: *Analyticity of the interface in a free boundary problem*, Math. Ann., 305, 435–459, (1996).

[31] J. ESCHER & G. SIMONETT: *A center manifold analysis for the Mullins-Sekerka model*, Journal of Differential Equations, **143** (1998), 267–292.

[32] A. FRIEDMAN : *Cancer models and their mathematical analysis*, Lect. Notes Math. , **1872**, 223–246 (2006).

[33] A. FRIEDMAN : *Mathematical analysis and challenges arising from models of tumor growth*, Math. Mod. Meth. Appl. Sci., **17**, 1751–1772 (2007).

[34] A. FRIEDMAN & G. LOLAS: *Analysis of a mathematical model of tumor lymphangiogenesis*, Math. Mod. Meth. Appl. Sci., **15**, 95–107 (2005).

[35] A. FRIEDMAN & F. REITICH: *Analysis of a mathematical model for the growth of tumors*, J. Math. Biol., **38**, 262–284 (1999).

[36] A. FRIEDMAN & F. REITICH: *Symmetry-breaking bifurcation of analytic solutions to free boundary problems*, Trans. Amer. Math. Soc., **353**, 1587–1634 (2001).

[37] D. GILBARG & T. S. TRUDINGER: " Elliptic Partial Differential Equations of Second Order", Springer–Verlag, New York, 2001.

[38] L. GRAZIANO & L. PREZIOSI: *Mechanics in tumor growth*, Modeling of Biological Materials, 263–322 (2006).

[39] F. P. GREENSPAN: *Models for the growth of a solid tumor by diffusion*, Stud. Appl. Math. , **52**, 317–340 (1972).

[40] F. P. GREENSPAN: *On the growth and stability of cell cultures and solid tumors*, J. Theor. Biol., **56**, 229–242 (1976).

[41] M. GYLLENBERG & G. WEBB: *A nonlinear structured population model of tumor growth with quiescence*, J. Math. Biol., **28**, 671–684 (1990).

[42] D. HANAHAN & R.A. WEINBERG: *The hallmarks of cancer*, Cell, **100**, 57-70 (2000).

[43] E. I. HANZAWA: *Classical solutions of the Stefan problem*, Tôhoku Math. J., **33**, 297–335 (1981).

[44] T. KATO: "Perturbation Theory for Linear Operators", Springer-Verlag, Berlin Heidelberg, 1995.

[45] A. LUNARDI: "Analytic Semigroups and Optimal Regularity in Parabolic Problems", Birkhäuser, Basel, 1995.

[46] W. RUDIN: "Functional Analysis" McGraw-Hill, New York, 1991.

[47] A. J. PERUMPANANI & H. M. BYRNE: *Extracellular matrix concentration exerts selection pressure on invasive cells*, Euro. j. Cancer, **35**, 1274–1280 (1999).

[48] A. J. PERUMPANANI, J. A. SHERRATT, J. NORBURY & H. M. BYRNE: *A two parameter family of traveling waves with a single barrier arising from the modelling of extracellular matrix mediated cellular invasion*, Phisica D, **126**, 145–159 (1999).

[49] G. SIMONETT: *Center manifolds for quasilinear reaction-diffusion systems*, Differential Integral Equations, **8**(4), 753–796 (1995).

[50] H. J. SCHMEISSER & H. TRIEBEL: "Topics in Fourier Analysis and Function Spaces", John Wiley and Sons Limited, New York, 1987.

[51] J. A. SHERRATT: *Traveling wave solutions of a mathematical model for tumor encapsulation*, SIAM J. Appl. Math., **60**, 392–407 (1999).

[52] J. A. SHERRATT & M. A. J. CHAPLAIN: *A new mathematical model for avascular tumor growth*, J. Math. Biol., **43**, 291–312 (2001).

[53] E. SINESTRARI: *On the abstract Cauchy problem of parabolic type in spaces of continuous functions*, Journal of Mathematical Analysis and Applications, **107**, 16–66 (1985).

[54] K. E. THOMPSON & H. M. BYRNE: *Modelling the internalization of labelled cells in tumor speroids*, J. Math. Biol, **61**, 601–623 (1999).

[55] S. TURNER & J. A. SHERRATT: *Intercellular adhesion and cancer invasion: A discrete simulation using the extended Potts model*, J. Theor. Biol., **216**, 85–100 (2002).

[56] F. VERULST: "Nonlinear Differential Equations and Dynamical Systems", Springer-Verlag, Berlin Heidelberg, 1990.

Die VDM Verlagsservicegesellschaft sucht für wissenschaftliche Verlage abgeschlossene und herausragende

Dissertationen, Habilitationen, Diplomarbeiten, Master Theses, Magisterarbeiten usw.

für die kostenlose Publikation als Fachbuch.

Sie verfügen über eine Arbeit, die hohen inhaltlichen und formalen Ansprüchen genügt, und haben Interesse an einer honorarvergüteten Publikation?

Dann senden Sie bitte erste Informationen über sich und Ihre Arbeit per Email an *info@vdm-vsg.de*.

Sie erhalten kurzfristig unser Feedback!

VDM Verlagsservicegesellschaft mbH
Dudweiler Landstr. 99
D - 66123 Saarbrücken

Telefon +49 681 3720 174
Fax   +49 681 3720 1749

www.vdm-vsg.de

Die VDM Verlagsservicegesellschaft mbH vertritt

Printed by Books on Demand GmbH, Norderstedt / Germany